X-Ray Absorption Spectroscopy for the Chemical and Materials Sciences

X-Ray Absorption Spectroscopy for the Chemical and Materials Sciences

John Evans

Professor Emeritus, University of Southampton, UK
Visiting Scientist, Diamond Light Source, UK

Registered Office(s)
John Wiley & Sons, Inc., 111 River Street, Hoboken, NJ 07030, USA
John Wiley & Sons Ltd, The Atrium, Southern Gate, Chichester, West Sussex, PO19 8SQ, UK

Editorial Office
The Atrium, Southern Gate, Chichester, West Sussex, PO19 8SQ, UK

For details of our global editorial offices, customer services, and more information about Wiley products visit us at www.wiley.com.

Wiley also publishes its books in a variety of electronic formats and by print-on-demand. Some content that appears in standard print versions of this book may not be available in other formats.

Library of Congress Cataloging-in-Publication Data

Names: Evans, John, 1949 June 2- author.
Title: X-ray absorption spectroscopy for the chemical and materials sciences / Professor, John Evans, Chemistry, University of Southampton, UK, Diamond Light Source, UK.
Description: First edition. | Hoboken, NJ : Wiley, [2018] | Includes bibliographical references and index. |
Identifiers: LCCN 2017024956 (print) | LCCN 2017027875 (ebook) | ISBN 9781118676172 (pdf) | ISBN 9781118676189 (epub) | ISBN 9781119990918 (hardback) | ISBN 9781119990901 (paperback)
Subjects: LCSH: X-ray spectroscopy.
Classification: LCC QD96.X2 (ebook) | LCC QD96.X2 E93 2018 (print) | DDC 543/.62–dc23
LC record available at https://lccn.loc.gov/2017024956

Cover Design: Wiley
Cover Image: Sound Waves on Water: © Sunny/Getty Images; Duck Images provided courtesy of John Evans

Set in 10/12pt Warnock by SPi Global, Pondicherry, India
Printed and bound in Singapore by Markono Print Media Pte Ltd

10 9 8 7 6 5 4 3 2 1

Contents

About the Author

John Evans hails from Newcastle upon Tyne. He studied Chemistry at Imperial College, London, and carried out his PhD at the University of Cambridge supervised by Lord (Jack) Lewis and Brian Johnson. His postdoctoral research was at Princeton University, with Jack Norton, and then with ICI and Royal Society Pickering Research Fellowships back at Cambridge. He moved with the Pickering Fellowship to Southampton in 1976, became a lecturer in 1978, and a professor in 1990. He is now an emeritus professor there. He was science program advisor at the Diamond Light Source Ltd from 2002 to 2007. His experience in applying XAFS spectroscopy to chemical problems extends over 35 years; his research group has carried out experiments at the SRS, ESRF, SLS, Hasylab, Diamond, and APS.

Preface

This is a textbook aimed at master's-level students, including fourth-year UK MSci degrees, of the chemical and related sciences suitable as an introductory text for PhD students embarking on x-ray absorption fine structure (XAFS) spectroscopy. The background should also appeal to established scientists from other fields (environmental, life, and engineering sciences), wishing to assess the potential of x-ray spectroscopy for their science. The chapters progress initially through the history and principles of XAFS. The next two chapters deal with experimental design: first, light sources and beamlines and then at the experimental station itself. Chapter 5 provides the background to the methods of extracting and using the results in materials and chemical analyses. The final chapter provides a series of case studies to illustrate a variety of applications. Each chapter concludes with a set of problems. There is a strong emphasis on the need to make the right choices for experimental design, and guidance provided to do so.

John Evans
Southampton UK
April 2017

Acknowledgments

I wish to thank all the members of my former research group for their talents and dedication in pursuing some optimistic experiments for 24/7 periods with food of varying desirability. Much of the developments came with collaborations that extended beyond a single position and with staff members from other institutions: Neville Greaves, Andy Dent, Sofia Diaz-Moreno, Norman Binsted, Trevor Gauntlett, Fred Mosselmans, Judith Corker, Steven Fiddy, Mark Newton, Moniek Tromp, Peter Wells, and Stuart Bartlett. Judith's loss to leukemia in 1998 remains a deep sadness. The book builds on the immense expertise of those who design, construct, develop, and operate these great accelerator-based light sources. Advances in science, technology medicine, and cultural heritage owe much to them.

In the writing of the book, I have been helped greatly by staff at Diamond and colleagues for providing raw data and graphics. Special thanks go to Stuart Bartlett, Andrew Hector, Fred Mosselmans, Sofia Diaz-Moreno, Roberto Boada Romero, Sarnjeet Dhesi, and Liz Duke. I am grateful, too, for the support of the CEOs of Diamond Light Source, Gerd Materlik, and Andrew Harrison, and also from EPSRC in the form of the Dynamic Structural Science and Catalysis Hub consortia at the Research Complex at Harwell. I am grateful for the confidence shown in this project by Jenny Cossham at Wiley and the continued patience of the staff at Wiley through the years. Inevitably, this has impacted on my family the most. Without the support of my wife, Hilary, and our daughters, Beccy and Lisa, and their families, this would not have reached fruition.

Glossary and Abbreviations

Absorption edge	Rapid increase in absorption with increasing energy
AEY	Auger electron yield
APD	Avalanche photodiode
Auger process	Relaxation of a core-hole via electron emission
CCD	Charge-coupled device
CEE	Constant emission energy
CIE	Constant incident energy
Compton scattering	Inelastic scattering
Debye-Waller	Factor describing disorder in interatomic distances
DFT	Density functional theory
EDE	Energy dispersive EXAFS
EDX	Energy dispersive x-ray spectroscopy
EXAFS	Extended x-ray absorption fine structure
FEL	Free electron laser
FT	Fourier transform
FY	Fluorescence yield
FZL	Fresnel zone plate
HARPES	Hard x-ray photoelectron spectroscopy
HERFD	High-energy resolution fluorescence detection
IV	In vacuum
KB	Kirkpatrick-Baez (mirrors)
MLL	Multilayer Laue lens
NEXAFS	Near-edge x-ray fine structure
NIXS	Nonresonant Inelastic x-ray Scattering
OD	Optically detected
PCA	Principal component analysis
QEXAFS	Quick extended x-ray absorption fine structure
Rayleigh scattering	Elastic scattering
REXS	Resonant x-ray Emission Spectroscopy
RIXS	Resonant Inelastic x-ray scattering or spectroscopy

SR	Synchrotron radiation
STXM	Scanning transmission x-ray microscopy
TEY	Total electron yield
TXM	Transmission x-ray microscopy
VtC	Valence to core
X-PEEM	X-ray photoelectron emission microscopy
XAFS	X-ray absorption fine structure
XANES	X-ray absorption near-edge structure
XAS	X-ray absorption spectroscopy
XEOL	X-ray excited optical luminescence
XES	X-ray emission spectroscopy
XFEL	X-ray free electron laser
XMCD	X-ray magnetic circular dichroism
XMLD	X-ray magnetic linear dichroism
XRS	(Inelastic) X-ray Raman Scattering

1

Introduction to X-Ray Absorption Fine Structure (XAFS)

1.1 Materials: Texture and Order

Today, research laboratories have powerful techniques for establishing the chemical nature and structure of pure materials. Our view of chemical structure is formed around the results of x-ray diffraction, recorded from single crystals or from polycrystalline powders. Structures in the liquid phase can be inferred from expectations for bond lengths and angles derived from crystallography; to do so, information is gathered about the local symmetry, atomic connectivity, and proximity in the material derived from structurally sensitive spectroscopies, particularly nuclear magnetic resonance (NMR) and infrared (IR) and Raman vibrational spectroscopies.

But many materials with a function are textured, such as pigments in paintings in the Louvre, a stained glass window in Westminster Abbey, an automotive exhaust catalyst, a dental filling, and others in nature, such as mineral inclusions or the shells of mollusks. They possess identifiable local structures on the Å scale that form the basis of their capabilities. However, these may be randomly spread through their three-dimensional shape or, alternatively, be located in a particular region, such as at a surface. Correlating the structure and the function of materials is a key to the design of further development, as well as providing its own intrinsic scientific elegance.

X-ray absorption fine structure (XAFS) spectroscopy has developed to the point when it can be applied to probe complex and faceted materials, for example, to reveal chromophores in glass and to probe the organic-inorganic composites in shells. In this book, the aim is to guide the readers to identify whether and how the technique might be used to advantage to study the materials that interests them within this wide spectrum of samples.

X-Ray Absorption Spectroscopy for the Chemical and Materials Sciences, First Edition. John Evans.
© 2018 John Wiley & Sons Ltd. Published 2018 by John Wiley & Sons Ltd.

1.2   Absorption and Emission of X-Rays

About 100 years ago, with the discovery of x-ray absorption (XAS) and emission (XES) spectroscopies, observation of the absorption and emission of x-rays were at the forefront of atomic physics, rather than the basis of materials characterization. The observations of the x-ray absorption edge of elements were first made by Maurice de Broglie in 1913 and published in 1916;[1] the elements were the silver and bromine in a photographic plate. Moseley[2] measured the energies of the emissions of over 40 elements and showed that there was a square root relationship with the atomic number of the element; tragically, his further contributions were cut short by a sniper at the Battle of Gallipoli in 1915. W.H. and W.L. Bragg had also noted that x-ray emission lines were also characteristic of an element.[3] Hence, both the absorption edge and the emission lines had been shown to provide a means of elemental speciation of sites.

Shortly thereafter the group of Manne Siegbahn at Lund improved the resolution of the crystal spectrometers to 1/10,000 allowing them to establish that the absorption edge position was chemically as well as elementally dependent; this was initially observed for allotropes of phosphorus, reported by Bergengren in 1920. In the next year, Lindh reported a chemical shift of 5.4 eV between Cl_2 and HCl. The use of edge positions for chemical speciation was thus established and by the mid-1920s the energies of emission lines were also shown to display a chemical shift.

1.3   XANES and EXAFS

In 1920, Fricke published photographic measurements of K absorption edges of elements between magnesium and chromium,[4] and Lindh reported structures around the Cl K edges. These reports showed fine structure both before and after the absorption edge energy, and XAFS (x-ray absorption fine structure) had been identified. Most photographic plates with the x-ray spectrum dispersed across them showed a bright line, marking the maximum in the x-ray absorption and thus little darkening of the photographic plate. For some samples, for example, the Ca K edge in calcite and gypsum, this feature was especially intense and by 1926, it was known as the white line.[5] Lindsay and van Dyke also reported features up to nearly 50 volts above the first main feature of the edge. Two years later, Nuttall[6] reported that the potassium K edges post-edge features could be used to distinguish between different minerals, and that the "fine structure… extended over a range of about 67 volts." And in 1930 Kievert and Lindsay[7] observed fine structures in metals extending to about 400 eV to higher energy of the absorption edge. Hence, by 1930 most of the core characteristics of XAFS spectroscopy had been identified, apart from polarization effects.

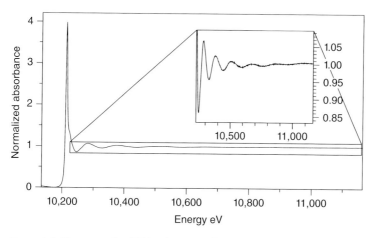

Figure 1.1 The normalized W L_3 edge x-ray absorption spectrum of a solution of $(NBu_4)_2[WO_4]$ (10 mM) in acetonitrile (*Source:* Diamond Light Source, B18; data from Richard Ilsley).

An example of an XAFS spectrum is shown in Figure 1.1 for the tungsten L_3 edge of an acetonitrile solution of $(NBu_4)_2[WO_4]$; the L_3 edge is a transition of a $2p$ electron of the absorbing atom, tungsten in this example. The technique pinpoints the anion containing the absorbing atom and the solvent and counter ion do not interfere. This spectrum shows some of the characteristic components that might be observed associated with an absorption edge. The x-ray absorption near-edge structure (XANES) is dominated in this case by an example of a white line, due to an intense (Laporte-allowed) transition to vacant $5d$ states. The extended x-ray absorption fine structure (EXAFS) has been expanded vertically to become visible at higher energies. Each of these types of features contributes to the information than can be derived from the entire spectrum.

1.4 Information Content

It was quickly recognized that x-ray spectra provided information about atomic energy levels, as commented by W.H. Bragg.[3] It was also noted that the position and shape of the XANES features at the absorption edge were dependent upon the local environment and on the effective charge on the absorbing atom. More problematical was a working explanation of the extended structure, EXAFS. There were three possibilities proposed:

1) The peaks above the edge were due to additional atomic transitions. However, Coster and van der Tuuk[8] showed that this was a minor contribution in their study on argon gas.

2) The oscillations were due to long-range periodicity through the sample, as described by Kronig in 1931.[9]
3) Instead the oscillations were due to short-range electron scattering, as Hartree, Kronig, and Petersen reported in 1934, thus accounting for EXAFS features up to 200 eV above the Ge K edge in molecular $GeCl_4$.[10]

The dichotomy between local- and long-range order explanations for solid-state arrays and molecular materials remained for about 40 years. The basis of the current understanding emanates from analyses by Stern and his co-workers, Sayers and Lytle, in 1970 and published in 1974–1975.[11–13] The key aspects of this development were the demonstration of the short-range order theory for all materials and the efficacy of Fourier transform methods for displaying the differing oscillations in an EXAFS patterns as distinguishable interatomic distances. In Figure 1.1, the EXAFS features are dominated by a single damped oscillation, which is due to the scattering between the tungsten and oxygen atoms in the anion. Hence the method provides measurement of that bond-length in solution and other disordered media.

1.5 Using X-Ray Sources as They Were

Viewing that oscillation in Figure 1.1, it is evident that EXAFS features are weak and thus a high signal/noise ratio is required to reliably extract the potential information in a XAS spectrum. Until 1970, all XAS measurements utilized laboratory x-ray tubes. For x-ray spectroscopy it is the brehmsstrahlung background that provides the necessary range of x-ray energies, rather than the more intense emission lines used for x-ray diffraction. The combined characteristics of weak sources and weak signals severely limited the application of XAFS. But the breakthrough in understanding provided by Stern added to the impetus for finding an experimental solution.

Much higher intensity sources were in prospect from synchrotron accelerators, an effect first demonstrated in 1947.[14] This report, from the General Electric Company, described a brilliant white spot emanating from the tangent point of the orbit in a 70 MeV device of radius 29.2 cm. When synchrotrons first became available as x-ray sources in the 1970s, the effect was dramatic. For example, the experimental backdrop to the theoretical developments was a suite of three x-ray spectrometers at the Boeing Scientific Research Laboratories. Lytle later offered the following observation[15] about an experimental trip to the then new x-ray spectrometer at Stanford Synchrotron Radiation Laboratory (SSRL) in the early 1970s: "In one trip to the synchrotron we collected more and better data in three days than in the previous ten years. I shut down all three X-ray spectrometers in the Boeing laboratory. A new era had arrived!"

1.6   Using Light Sources Now and To Be

That new era transformed x-ray spectroscopy from being a poorly understood technique of considerable experimental challenge, to one of wide applicability. Stern, Sayers, and Lytle concluded in 1975:[13] "Its greatest usefulness should be in unraveling complicated structures with no long-range order such as biological molecules and commercially practical catalysts."

Since that statement was made, x-ray sources display brilliances that have increased faster than computing power over the same time period. The advantages envisaged then for studies of macroscopic samples can now be applied with great spatial and temporal resolution. Functional materials may be investigated in their textured nature and their structures tracked during processing. Current storage ring light sources have the stability and reliability to make these experiments viable.

A new class of x-ray sources is within the horizon now. Lasing by free electron lasers in the x-ray region was demonstrated in the United States in 2010 and in Japan in 2011, and both are available as user facilities now. This type of source will be able to interrogate structure with a time resolution that is faster than vibrations and thus they offer structural snapshots of molecular dynamics. The last century has been a good one for x-ray science, and the future is extraordinarily bright!

For now though, this is what we hope to probe by XAFS experiments:

- The absorption edge jump and fluorescence yield can be used to quantify elemental compositions.
- The XANES features and observed edge position interrogate the effective nuclear charge of the absorbing atom and the local geometry.
- The oscillations in the EXAFS region provide information about the types of neighboring atom, the number of them, and the interatomic distances.
- Spatial resolutions of μm down to 10s of nm provide scope for using the spectral features for mapping chemical states though textured materials.
- Sub-second time resolutions allow tracking of structural changes in response to a stimulus, thus giving real-time structure-function studies.
- With specialized laser-based x-ray sources, structures can be monitored with sub picosecond time resolution.

The events involved can be envisaged with a physical model. In Figure 1.2, an energy source (a stone) is exciting a calm lake, creating a core hole and also a wave from the impact site.

In XAFS, there are generally neighboring atoms as well as the excitation site, so perhaps a better model is provided by a coot (*fulica atra*). Coots dive like a stone to feed and this "excited state" has a central void with an outgoing wave. The wave can interact with a nearby coot and the "excited" coot relaxes (Figure 1.3). A portion of the outgoing wave can be scattered back from the hungry coot back to the one that has just fed.

(a)

(b)

(c)

Figure 1.2 Successive photographs of a stone entering a lake depicting initial excitation, hole formation, and wave development.

(a)

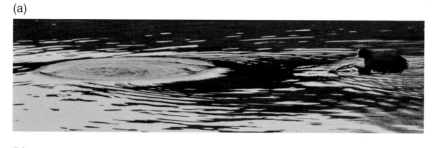

(b)

Figure 1.3 The hole and wave created by a diving coot (top) and the interaction of the wave with a neighbor (bottom).

1.7   Questions

Find the 1975 paper of Lytle, Sayers, and Stern on the experimental practice of EXAFS.[12]

1   Read the section on the experimental apparatus. It gives a detailed description from carrying out a measurement to extracting the EXAFS data from the resulting spectrum in Figure 3 in the 1975 paper.

2   Note the effect on the temperature on the EXAFS spectrum of copper (Figure 4 in the 1975 paper) and account for it.
   A   Comparing this with the three spectra in Figure 5 in the 1975 paper, with which metal does it share the same structure?
   B   Figure 6 in the 1975 paper shows different spectral profiles for different neighbors for germanium. How does this help identify the type of neighboring atom?
   C   Account for the appearance of the spectrum of ferrocene in Figure 8 in the 1975 paper.

References

1 'La spectrographie des phénomènes d'absorption des rayons X', M. De Broglie, J. Phys. Theor. Appl., 1916, **6**, 161–168.

2 'The high frequency spectra of the elements', H. G. J. Moseley, Philos. Mag., 1913, **26**, 1024–1034; 1914, **27**, 703–713.

3 'The reflection of x-rays by crystals. II', W.H. Bragg, Proc. R. Soc., Lond. A, 1913, **89**, 246–248.

4 'The characteristic absorption frequencies for the chemical elements magnesium to chromium', H. Fricke, Phys. Rev., 1920, **16**, 202–216.

5 'The K x-ray absorption of calcium in calcite, gypsum and fluorite', G. A. Lindsay, G. D. van Dyke, Phys. Rev., 1926, **28**, 613–619.

6 'The K absorption edges of potassium and chlorine in various compounds', J. M. Nuttall, Phys. Rev., 1928, **31**, 742–747.

7 'Fine structure in the x-ray absorption spectra of the K series of the elements calcium to gallium', B. Kievit, G. A. Lindsay, Phys. Rev., 1930, **36**, 648–664.

8 'The fine structure of the x-ray absorption edge in the K-series of argon and its possible interpretations', D. Coster, J. H. van der Tuuk, Nature, 1926, **117**, 586–587.

9 a'Zur Theorie der Feinstruktur in den Röntgenabsorptionsspektren', R. de L. Kronig, Z. Phys. 1931, **70**, 317–323; b'Quantum mechanics of electrons in crystal lattices', R. de L. Kronig, W.G. Penney, Proc. R. Soc., Lond. A, 1931, **130**, 499–513.

10 'A theoretical calculation of the fine structure for the *K*-absorption band of Ge in GeCl₄', D. R. Hartree, R. de L. Kronig, H. Petersen, Physica, 1934, **1**, 895–924.

11 'Theory of extended x-ray absorption fine structure', E. A. Stern, Phys. Rev. B, 1974, **10**, 3027–3037.

12 'Extended x-ray absorption technique. II. Experimental practice and selected results', F. W. Lytle, D. E. Sayers, E. A. Stern, Phys. Rev. B, 1975, **11**, 4825–4835.

13 'Extended x-ray absorption technique. III. Determination of physical parameters', E. A. Stern, D. E. Sayers, F. W. Lytle, Phys. Rev. B, 1975, **11**, 4836–4846.

14 'Radiation from electrons in a synchrotron', F. R. Elder, A. M. Gurewitsch, R. V. Langmuir, H. M. Pollock, Phys. Rev., 1947, **71**, 829–830.

15 'The EXAFS family tree: a personal history of the development of x-ray absorption fine structure', F. W. Lytle, J. Synchrotron Radiat., 1999, **6**, 123–134.

2

Basis of XAFS

2.1 Interactions of X-Rays With Matter

When electromagnetic waves of x-ray energies (>120 eV, wavelength < 10 nm) interact with the electron clouds of a material, three principal effects can dominate the result (Figure 2.1).

1) There is elastic (Rayleigh) scattering with the incoming and outgoing photons having the same energy. Diffraction is a specific case of this for crystalline samples.
2) Additionally, there is inelastic (Compton) scattering. In this case the outgoing photon is of lower energy (longer wavelength). This is similar to Raman spectroscopy, which generally employs photons in the visible or near infrared to probe molecular vibrations or rotations. Using x-rays as the excitations, the energy transfer can be to vibrations and to electronic transitions, both core and valence. Hence this can include components of XAFS spectroscopy.
3) For x-ray absorption spectroscopy the most important type of event is absorption. This photoelectric effect involves a transition of an electron in a core orbital creating an excited state. This highly excited state can relax by emitting energetic electrons and photons of longer wavelength (fluorescence).

2.1.1 Absorption Coefficients

As shown in Figure 2.1, absorption attenuates the x-ray flux transmitted through a sample, which follows an exponential reduction by:

$$I_t \big/ I_0 = e^{-\mu l} \tag{2.1}$$

where I_0 and I_t are the incident and transmitted intensities, respectively, μ is the linear absorption coefficient, and l the sample length. The mass absorption coefficients, μ_m (=μ/ρ, where ρ = the density of the material) are tabulated for all elements[1–3] and thus the absorption characteristics of any material can

X-Ray Absorption Spectroscopy for the Chemical and Materials Sciences, First Edition. John Evans.
© 2018 John Wiley & Sons Ltd. Published 2018 by John Wiley & Sons Ltd.

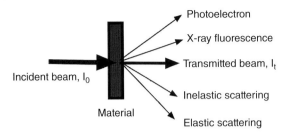

Figure 2.1 Interaction of x-rays with materials.

be calculated by a mass-weighted summation of the absorption coefficients of the elements at a given energy, according to equation 2.2.

$$log_e\left(I_t/I_0\right) = \Sigma_i w_i l_i \left(\mu_m/\rho\right)_i \qquad (2.2)$$

The absorption properties of elements vary greatly with atomic number. Higher Z elements have more electrons at deeper potentials and thus have stronger interactions with incoming x-rays. As the energy of x-rays increase then the interaction becomes weaker and transmission increases through a sample. This is a strong effect related approximately to the λ^3. The effect can be seen in Figure 2.2. This shows that the absorption of a solution of a salt of the tungsten oxo-anion in a solvent of light elements decreases very significantly even over the energy range of a single XAFS spectrum.

2.1.2 Absorption Edges

As is evident in Figure 2.2, the drop in absorption coefficient is arrested by a sharp rise in absorption, termed the absorption edge. This occurs when the photon energy corresponds to the binding energy of a core orbital. Absorption edges are conventionally labeled alphabetically according to the principal quantum number of the electron shell of the element being studied: K for $n=1$, L for $n=2$, M, for $n=3$, and so on. Hence the label, K, corresponds unambiguously to a transition from a $1s$ orbital. For the second shell, different energies would be anticipated for excitation of the $2s$ and $2p$ orbitals. Accordingly, the next highest absorption edge energy after the K is attributed to the $2s$ orbital, and is termed variously the $L(I)$, L_I or L_1 edge. However, rather than being a single absorption edge resulting from the $2p$ electrons there are two—the L_2 and L_3 edges.

The removal of an electron by photo-ionization creates an electron vacancy (a core hole) leaving a $2p^5$ sub-shell. Coupling between the orbital angular momentum, l (1 for a p orbital), and the electron spin, m_s (½), gives rise to two j states: $2p_{3/2}$ and $2p_{1/2}$. In accord with Hund's rules, the former, corresponding

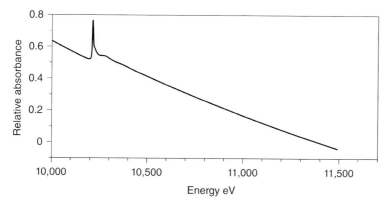

Figure 2.2 W L_3 XAS of $(NBu_4)_2[WO_4]$ (10 mM) in CH_3CN (*Source:* Diamond, B18, data from Richard Ilsley).

to $j = l + s$, should be the more stable ion with the sub-shell being more than half full. So the third highest energy edge will be the L_2 with the ion having a $2p_{1/2}$ sub-shell and the fourth will be the L_3 edge affording a $2p_{3/2}$ configuration. Figure 2.2 is showing the L_3 edge of tungsten, with an energy near 10207 eV. At higher energies lie the K edge at 69525 eV, the L_1 at 12100 eV, and the L_2 at 11544 eV, just past the end of the spectrum in Figure 2.2. These values show the large difference in energy between the electrons in the $1s$ and $2s$ orbitals, and a smaller difference, but still substantial, between $2s$ and $2p$. Rather less obvious, perhaps, is the large spin-orbit coupling energy giving a 1300 eV difference between the L_2 and L_3 edges. This is due to the strong relativistic effects of these core electrons, which have very high kinetic energies. This absorption edge was studied for an experiment in preference to the L_2 edge for two reasons.

1) It is the more intense. This is due to the higher degeneracy of the $j = 3/2$ state as compared to the ½ (4 versus 2).
2) From the energy of the L_3 edge up to that of the L_2 all features in the spectrum are unambiguously associated with the L_3 transition. After the L_2 edge, there will be an overlap of features from the two edges. Due to the large energy difference between these two particular edges, this effect will be small, but that might not be the case for the corresponding edges of molybdenum, the element above tungsten in Group 6, which differ by about 100 eV (L_3 2520 eV, L_2 2625 eV).

The same principle applies to edges based upon $3d$ orbitals. There will be two j states generated—$3d_{3/2}$ and $3d_{5/2}$—with the latter being the more less stable and thus requiring less energy to create. This is summarized in Table 2.1.

Tables of x-ray absorption energies are readily available, not the least from the "X-ray Data Booklet."[3]

2.1.3 XANES and EXAFS

As indicated above, the absorption edge is associated with transitions of electrons in core orbitals to vacant states. Two types of vacant state can be envisaged: i) to a vacant valence orbital and ii) leaving the absorbing atom as a wave into the continuum (Figure 2.3). The first of these (Figure 2.3a) will affect the features observed near the absorption edge threshold, including pre-edge features. The virtual valence orbitals accepting the electron promoted from the core may be either largely localized on the absorbing atom, or one that is very substantially delocalized onto neighboring atoms, and thus have charge transfer characteristics. In Figure 2.3b, the x-ray is depicted as having sufficient

Table 2.1 Absorption edges and their corresponding electron configurations.

Absorption edge	Electron excited	Excited states
K	1s	$1s_{1/2}$
L_1	2s	$2s_{1/2}$
L_2, L_3	2p	$2p_{1/2}, 2p_{/3/2}$
M_1	3s	$3s_{1/2}$
M_2, M_3	3p	$3p_{1/2}, 3p_{/3/2}$
M_4, M_5	3d	$3d_{3/2}, 3d_{/5/2}$

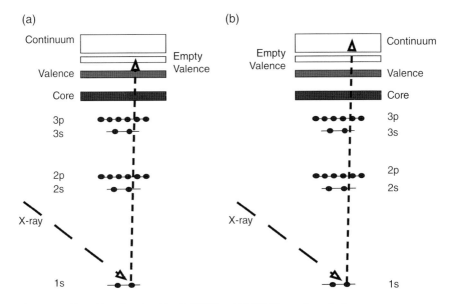

Figure 2.3 Transitions involved in a) XANES and b) EXAFS spectrum features.

energy to escape from the absorbing atom, and thus have an excess kinetic energy. This kinetic energy provides the photoelectron with its momentum and wavelength.

2.1.3.1 XANES Features

The features in the XANES region then are due to transitions to unoccupied (virtual) states. The most probable pathway will involve dipole (Laporte) allowed transitions, which follow the selection rule $\Delta l = \pm 1$. Thus for K absorption edges, where the ground state electron is in the $1s$ orbital ($l = 0$), it is the density of the empty p states that is most important. This will also hold for the less studied L_1 edge, which involves transitions from the $2s$ orbital. For the L_3 edge, transitions to the both s and d states are formally allowed, but those to the d states have a substantially higher probability. The effects of this are illustrated in Figure 2.4, which compares the L_3 and L_1 edges of a tetrahedral ion with a $5d^0$ valence configuration.

As is evident, there is a strongly allowed absorption at the absorption edge at the L_3 edge. This can be ascribed to a dipole allowed transition of an electron from the $2p$ orbital to the $5d$ states of tungsten. All of these are vacant providing a high probability for the transitions, which, in this anion, are symmetry allowed. The lower energy set of these, of e symmetry in the tetrahedral point group T_d, (Figure 2.5a) displays considerable $5d$ character, but there is much $5d$-$6p$ mixing evident in the t_2 set (Figure 2.5b). However, at the L_1 edge, an allowed transition from the $2s$ orbital (a_1 symmetry) would be have t_2 symmetry, as in that shown in Figure 2.5b. There is a prominent absorption on the rising slope of the absorption edge. This arises from the significant degree of allowed-ness in this transition, but the probability is not as high as for the totally allowed $2p$ to $5d$ transitions as seen in the L_3 edge.

The hybridization effect would be expected to be dependent upon the geometry and symmetry at the absorbing atom. In Figure 2.6, the K edge XANES features of a selection of chromium centers are presented. Here the tetrahedral complex CrO_4^{2-} displays an intense first absorption that is significantly below the energy of the absorption edge (14 eV), forming this very strong pre-edge feature. This Cr(VI), complex will also have a $e^0 t_2^0$ valence configuration with the d-p hybridized orbital set (t_2) unfilled. One of these triply degenerate orbitals is illustrated in Figure 2.7; the p-d mixing is high in this case, with the p character predominating. Also shown in this figure is one of the vacant triplet of β-spin states associated with the t_{2g}^3 configuration of $[Cr(OH_2)_6]^{3+}$. In a centrosymmetric site, p-d mixing is symmetry forbidden, and thus the low-lying empty states are essentially $3d$ in character. Thus transitions from a $1s$ orbital to one of these states will require a different mechanism such as distortion by coupling of the transition with an antisymmetric vibration of the metal center (e.g., the t_{1u} antisymmetric O-Cr-O stretch) or via a higher order (quadrupole) transition. Both of these effects will be of lower probability and the resulting

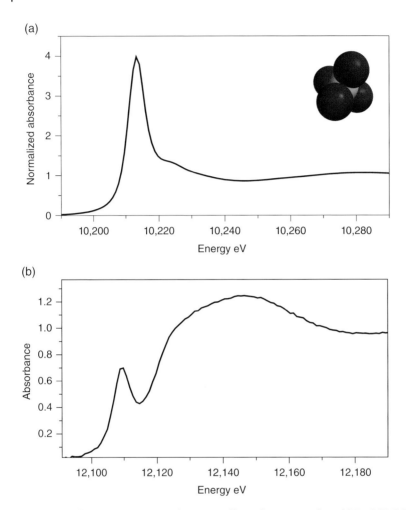

Figure 2.4 The XANES regions at the a) L_3 and b) L_1 absorption edge of (NBu$_4$)$_2$[WO$_4$] (inset) (10 mM) in CH$_3$CN solution (10 mM) recorded in transmission (*Source:* Diamond, B18, data from Richard Ilsley).

transitions will be weaker. Hence, the octahedral centers in [Cr(CO)$_6$] ($t_{2g}^6 e_g^0$) and Cr$_2$O$_3$ ($t_{2g}^3 e_g^0$), which both also have vacant *3d* levels of different symmetry to that of the vacant 4p orbitals (t_{1u}) display weaker pre-edge features.

The edge positions of Cr metal, Cr^{III}$_2$O$_3$, and K$_2$[Cr^{VI}O$_4$] follow a trend with oxidation state: an increase in Z_{eff} with oxidation state, which will lower the energy of the *1s* core orbital and will increase the ionization potential (Section 2.1.3.2). However, [Cr^0(CO)$_6$] exhibits an absorption edge energy very

(a) (b)

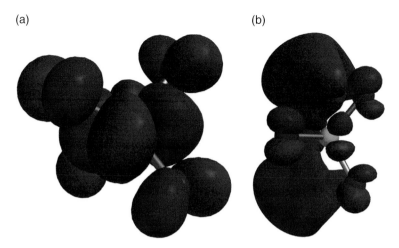

Figure 2.5 Low-lying vacant states as calculated for [WO$_4$]$^{2-}$ in aqueous solution using Spartan'16 using the ωB97X-D/6-31G* method. a) one of the *e* set and b) one of the *t$_2$* set.

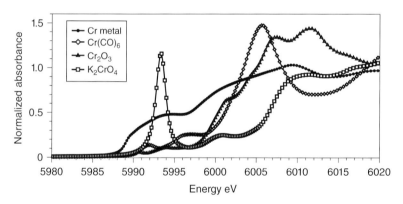

Figure 2.6 Cr *K* edge XANES of Cr metal, [Cr^0(CO)$_6$], Cr$^{III}_2$O$_3$ and K$_2$[Cr^{VI}O$_4$] (*Source:* Data courtesy of Sofia Diaz-Moreno, Roberto Boada-Romero, and Luke Keenan, Diamond, I20).

similar to that of the Cr^{III} oxide. This demonstrates that the position of the absorption edge is also a function of the ligand set, and indeed upon the coordination geometry, particularly for soft-donor ligands.[4]

Near to, and above, the energy of the absorption edge the excited electron has attained the Fermi energy and can move within the material. At low kinetic energies (<10 eV) the mean free path of the photoelectron in solids is relatively long (1 to 10 nm) (Figure 2.8). It falls to a minimum of about 0.5 nm at 50–100 eV, and only rising above 1 nm again above 1000 eV. Hence, in the XANES region, the photoelectron can interrogate and be scattered by structural arrays of a significant number of shells. In principle these states might be calculated and

(a)   (b)

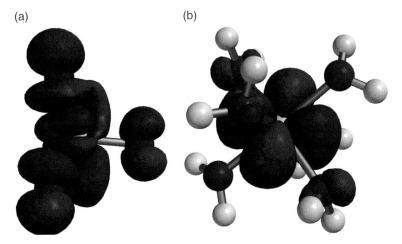

Figure 2.7 Low-lying vacant states as calculated for two chromium complexes in aqueous solution using Spartan'16 using the ωB97X-D/6-31G* method. a) $[CrO_4]^{2-}$ and b) $[Cr(OH_2)_6]^{3+}$.

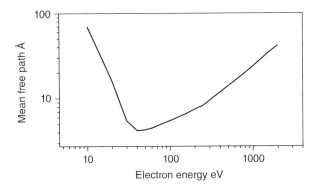

Figure 2.8 Typical electron mean free path in a material—modeled with silicon.

discussed in a quantum mechanical way; the large scale of the slab of the structure that may need to be considered can make this challenging (Section 5.3.2). In the XANES region, the photoelectron electron has such a long mean free path that the scattering is from a large number of atoms—called *multiple scattering*. XANES features can then be challenging to model and interpret and so have more often been used in a finger print mode in comparison with reference materials of known structure.

2.1.3.2 Edge Position

The energy of the absorption edge itself had been identified as chemically dependent early in the history of x-ray absorption spectroscopy (Section 1.2). The apparent position of the absorption edge can be seen to occur at different energies in Figure 2.6.

There have been two alternative definitions used for measuring the edge position:

1) The energy of the absorption edge at 50% of the edge jump and
2) The energy of maximum slope (first derivative) on the absorption edge.

The latter is the more commonly employed measurement point. It is a directly observable point whereas the edge jump needs to be estimated by a backward extrapolation from the post-edge spectrum. The maximum slope is close to, but a few eV above, the Fermi energy of a metal. It thus provides a starting estimation of the threshold energy of the photoelectron, E_0.

The absorption edge energy then may be related to the binding energy of photo-excited electron, which will be affected by the Z_{eff} (Effective Atomic Number) of the binding site. If Z_{eff} is increased, for example, by increasing the oxidation number of the absorbing atom, the overall charge on a complex or increasing the electronegativity of neighboring atoms, then the energy levels of the core orbitals would be lowered, thus increasing their binding energies (Figure 2.9). In an excited state, with the electron transferred into an unoccupied valence orbital, then an increased Z_{eff} will also increase the binding energy of the destination orbital. The edge shift will be the difference between these two values, hence the edge position will increase with Z_{eff} if

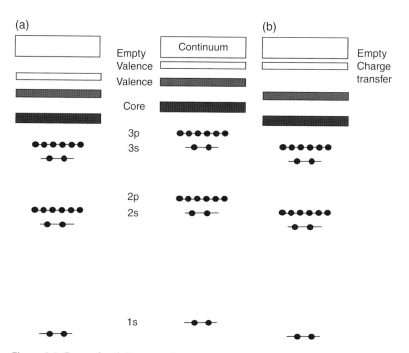

Figure 2.9 Energy level diagrams showing the effect of increasing the Z_{eff} from that of the reference material (center): a) when the transition is localized on the absorbing atom and b) if there is significant charge transfer.

the effect on the core orbital dominates. If the transition is largely charge transfer in nature, then the binding energy of the acceptor orbital would not be expected to change much with the effective atomic number of the absorbing atom, and in this case the edge position will certainly be dominated by the ground state effects.

Experimental evidence for the usefulness of the edge position can be seen from sulfur K edge spectra (Figure 2.10). The apparent edge position varies by about 10 eV between sulfide, S^{2-}, and sulfate(VI), SO_4^{2-}. This follows the direction expected if the effect to be dominated by the shift in the binding energy of the core orbital. The $1s$ orbital is relatively shallow in energy and screening by the $3s$ and $3p$ valance electrons of sulfur would be expected to cause a significant effect. So differentiation of S(-II), S(0), S(IV), and S(VI) is viable from such measurements. Interesting too is the increase in the height of the first transition from a $3s^2 3p^6$ configuration through to $3s^0 3p^0$. The increase in p vacancies (holes) (n_h) enhances the intensity of the dipole allowed $1s$ to $3p$ transition. For both $[SO_4]^{2-}$ and $[SO_3]^{2-}$ the acceptor orbitals are largely of S $3p$ character (Figure 2.11) and are S-O antibonding in nature; for sulfate(VI), this is triply degenerate as compared to being doubly degenerate for sulfate(IV). On that ground alone, the pre-edge feature would be anticipated to be more intense for $[SO_4]^{2-}$. The intensity of both these features masks the onset of the continuum states normally considered to represent the absorption edge.

However, as we can imagine from the spectra shown in the figures above, the measured-edge position is a function of the perceived Z_{eff}, the coordination

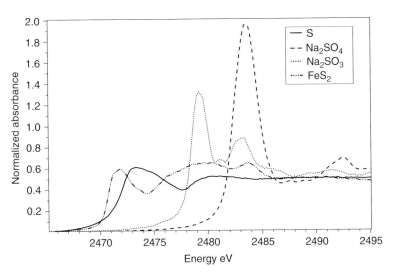

Figure 2.10 Sulfur K edge XANES of a series of compounds of different oxidation state (*Source*: Recorded on Lucia when at the Swiss Light Source, data from Michal Perdjon-Abel).

(a)   (b)

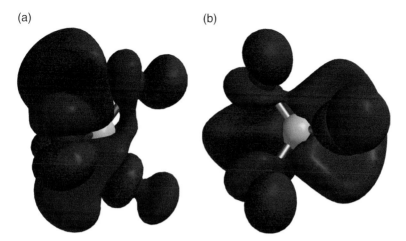

Figure 2.11 Low-lying vacant states as calculated for two sulfate ions in aqueous solution using Spartan'16 using the ωB97X-D/6-31G* method. a) $[SO_4]^{2-}$ and b) $[SO_3]^{2-}$.

geometry, and the valence electron configuration. So correlations between edge position and oxidation number must be carried out judiciously. As is evident from the chromium examples in Figure 2.6, the nature of the ligand as well as the oxidation state and coordination geometry can have a significant influence.

2.1.3.3 The EXAFS Effect

From Chapter 1, we can see that less intense, broader oscillations than those in the XANES region can extend for 100s of eVs beyond the absorption edge. In the energy range of 50 to 1000 eV, the mean free path of the photoelectron is in the range of 5–10 Å. The cause of the oscillation should then be local, and it is described by back-scattering by the electron clouds from neighboring atoms (Figure 2.12).

The photoelectron behaves as a wave the wavelength (λ). This wavelength is related to its momentum (p) by the de Broglie equation $\lambda = h/p$, where h is Planck's constant. The momentum is also related to the *photo-electron wave vector*, k, by $p = \hbar k$, with $\hbar = h/2\pi$. Since kinetic energy and momentum are defined as $E = \frac{1}{2}mv^2$ and $p = m_e v$, the following relationships emerge for an electron of mass m_e: $2Em_e = p^2$ and thus $\hbar k = \sqrt{(2Em_e)}$. At a particular X-ray photon energy E above the threshold energy Eo, the photoelectron wave vector, of units Å^{-1}, will be given by:

$$k = \sqrt{2m_e \left(E - E_o \right) / \hbar^2} \tag{2.3}$$

$$= \sqrt{0.262449 \left(E - E_o \right)} \tag{2.4}$$

(a)

(b)

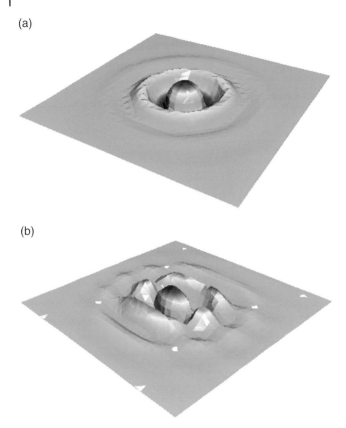

Figure 2.12 a) The photoelectron outgoing wave and b) with back scattering from a neighboring atom.

The electron wave has the wavelength, $\lambda = 2\pi/k$. The effect of the backscattered wave is that when it is in phase with the photoelectron wave the probability of absorption increases; it is decreased when these waves are out of phase. The result is that the absorption has a sinusoidal relationship to the wave vector k.

EXAFS datasets are based upon fractional changes in absorption as a function of k, $\chi(k)$. This fraction is the deviation between the observed absorption jump above E_0 from the expected jump if there were no EXAFS features. This can be written as:

$$\chi(k) = \frac{\mu(k) - \mu_0(k)}{\mu_0(k)} \tag{2.5}$$

As a result, $\chi(k)$ requires identifying E_0 from the observed spectrum, extrapolating the absorption through edge as if the edge were not there (pre-edge background), and identifying the absorption post edge as if there were no EXAFS (post-edge background). The background edge jump is the difference between the two backgrounds, $\mu_0(k)$. The values $\mu(k)$ are from the difference between the pre-edge background and the observed absorption. The steps required are illustrated in Figure 2.13.

As can be seen, the EXAFS oscillations decay quickly with increasing k value. In this example, the back-scattering is from a single type of oxygen atoms within the tetrahedral $[WO_4]^{2-}$ ion. In order to show the oscillations more clearly across the spectrum, the XAFS amplitudes that are viewed and analyzed are generally multiplied by k^n, where n is generally 1 to 3; k^2 weighting illustrated in Figure 2.13. The Fourier transform of the last curve shows a dominant Fourier component in real space (R in Å) with a distance similar to, but shorter than, the expected value for the W-O bond length.

2.1.3.4 EXAFS Quantification

The local scattering model proposed by Stern was a key to understanding the application of EXAFS in the analysis (Section 1.4). The model was based upon a single scattering process, in other words the photoelectron is considered as being scattered back from one atom, thus traveling $2R_j$, for an interatomic distance R_j. The photoelectron wave is modeled as a plane wave interacting with small atoms, approximations that hold more closely in the EXAFS region with photoelectron energies of more than 50 eV. This was known to be a practical approach. As will be discussed in Chapter 5, full analysis relaxes these approximations with the electron wave considered to be a sum of spherical harmonics and higher order scattering, in which the photo-electron visits two or more atoms before its return to the central atom. Equation 2.6, which applies to a randomly oriented sample like a powder, glass or solution, provides a good basis to introduce the factors important in EXAFS.

$$\chi(k) = \sum_j S_o^2 N_j \frac{|f_j(k)|}{kR_j^2} \sin(2kR_j + 2\delta_c + \phi) e^{-2R_j/\lambda(k)} e^{-2\sigma_j^2 k^2} \tag{2.6}$$

The observed EXAFS, $\chi(k)$, is a summation of the back-scattering for a series of shells each composed of the same atom within a relatively narrow band of distances. Experimentally, one would like to determine the interatomic distance between the central absorbing atom and the scattering atom (R_j). The sinusoidal term is dominated by this factor and thus the oscillations provide a good measure of the interatomic distance. The second important information element is the number of atoms in that shell, N_j. The amplitudes of the EXAFS features are linearly related to the magnitude of the coordination number. The third piece of information would be to identify the element within that shell. This does not appear to be within equation 2.6 at first sight, but there are two terms that bear the characteristics of the back-scattering element. One is the

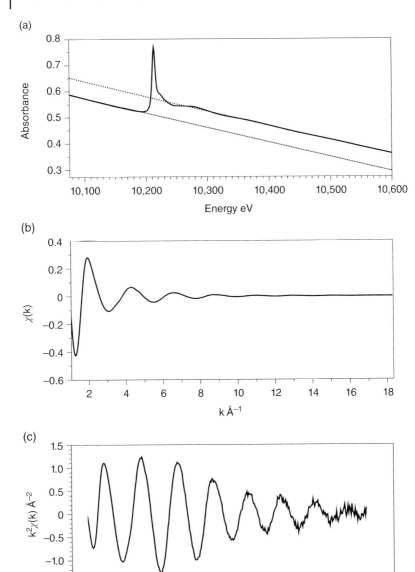

Figure 2.13 Steps to identify EXAFS, $\chi(k)$. a) Experimental x-ray absorption spectrum of the L_3 edge of $(NBu_4)_2[WO_4]$ in MeCN solution, showing pre-edge background and post-edge background, b) the resulting $\chi(k)$, c) the k^2 weighted EXAFS, $k^2.\chi(k)$ and d) the magnitude (solid), real part (dashed) and imaginary part (dotted) of the Fourier transform of $k^2.\chi(k)$ (*Source*: Diamond, B18, data from Richard Ilsley).

(d)

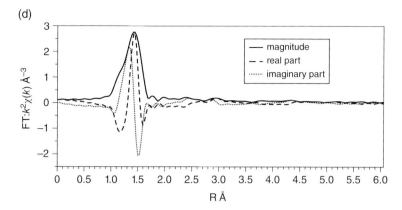

Figure 2.13 (Cont'd)

back-scattering amplitude *f(k)*. This provides the shape envelop of the back-scattering as a function of *k*. The amplitude is related to the phase shift experienced by the photoelectron as it encounters the potential of the back-scattering atom. The other term within the sin function is the phase shift of the central atom, δ_c, which is experienced twice: when leaving and returning to the central atom. These atomic factors used to be estimated empirically, but now are generally calculated reliably. They can therefore be used to fingerprint the row of the periodic table of the back-scattering atom.

The other terms in equation 2.6 generally restrict the range and precision of the information about the structure around the absorbing atom. The range is restricted as the photoelectron wave will become less dense with increased distance (by R_j^2). Also, the mean free path of the photoelectron, $\lambda(k)$, will also ensure that the information observed in the EXAFS region is local to less than 1 nm. The precision of the coordination number, N_j, is compromised by two other amplitude related factors in equation 2.6. The factor S_o^2 allows for the fact that not all of the absorbed x-rays lead to single electron process and therefore to EXAFS. In practice this value is about 0.8 ± 0.1, and this alone can result in a 10% margin of error in the coordination number. The last term, an exponential, contains the Debye-Waller factor, σ (Å). Based upon there being a Gaussian distribution of interatomic distance within a shell, this estimates the spread of the distribution. This disorder can be either *static*, due to an intrinsic variation within the structure, or *dynamic*, due to the spread of distance caused by thermal vibrations. The timescale of the x-ray transition is much faster than the timescale of a vibration and so XAFS probes the distributions of the interatomic distances in the vibrational envelope. This factor will reduce with lowered sample temperatures. The *k* dependence of this exponential means that an increased Debye-Waller factor will dampen the EXAFS especially at higher *k*. Not only might that make it difficult to unravel estimate the coordination number, it will also reduce the accuracy of the inter-atomic distance, R_j, by reducing the range of the useful data.

The effects of changes in R_j and the Debye-Waller factor, can be envisaged using equation 2.7, an extract from equation 2.6.

$$\chi(k) = \frac{1}{k} N \sin(2kR) e^{-2\sigma_j^2 k^2} \tag{2.7}$$

In Figure 2.14a, the effect of changing the interatomic distance is shown. When the distance is increased the k interval for an oscillation is decreased. In different parts of the spectrum the two waves vary between being nearly completely out of phase to being in phase, and a beat will result. The plot shows the value of having a large k range to distinguish these two distances accurately. The effect of increasing the Debye-Waller factor is shown in Figure 2.14b. The damping of the oscillation is increased and this is more serious at higher k values, effectively truncating the useable k range.

This word of warning should not be taken as a counsel of despair. EXAFS can provide significant structural information about the 5–6 Å space around an absorbing atom. As an example, the W L_3 EXAFS pattern of the cluster anion, $[W_6O_{19}]^{2-}$, shows complicated patterns indicative of contributions from many shells (Figure 2.15a). As well as the strong features below $k = 9 \text{ Å}^{-1}$ attributable in significant part to W-O back-scattering, there is clear evidence of another envelop peaking near 16 Å^{-1}. This is characteristic of back-scattering from a high Z element, tungsten in this case. The amplitude of the Fourier transform of this pattern (Figure 2.15b) shows two strong components due to the terminal and edge-bridging O atoms. The distances in the Fourier transform are shorter than the expected bond lengths due to the effect of the phase shift terms in the sin function in equation 2.6. There are strong contributions too from W-O-*cis*-W back-scattering, even though these are non-bonded interactions. The highest R contributions apparent result from back-scattering from the W-O-*trans*-W unit, which is about 4.6 Å distant. These are relatively long distances to observe in the EXAFS of a molecular ion, especially with no metal-metal bonding. This is due to a combination of the high overall symmetry, the high atomic number of the non-bonded back-scatterers, and the relatively low Debye-Waller factors created by the cage structure.

2.2   Secondary Emissions

The excited state created by the absorption of the x-ray has a short lifetime. Indeed, core-hole lifetimes (τ) can afford very significant Heisenberg uncertainty broadening (Γ) limiting the resolution of XANES features (equation 2.8); this effect increases for deeper core holes. For example, the broadening at the sulfur K edge at 2.47 keV is rather modest at 0.59 eV, as it is for the Fe K edge (1.25 eV at 7.11 keV). However, for lead the line broadening becomes very

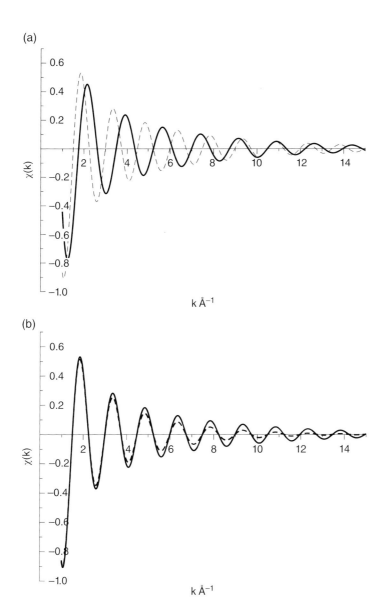

Figure 2.14 a) The oscillations calculated for an interatomic distance, R, of 1.8 (solid line) and 2.1 Å (dashed line), taking N as 1 and $2\sigma^2$ as 0.005 Å^2. b) Oscillations calculated for $2\sigma^2$ as 0.005 (solid line) and 0.015 Å^2 (dashed line), taking N as 1 and R as 2.1 Å.

large: 60 eV for the K edge at 88 keV, and would broaden EXAFS as well as XANES. This is one reason why the preferred edge for lead is the L_3 (13 keV) with a broadening of 5.8 eV. The decay of the core-hole is accompanied by energy loss either by electron excitation or the emission of a fluorescent photon.

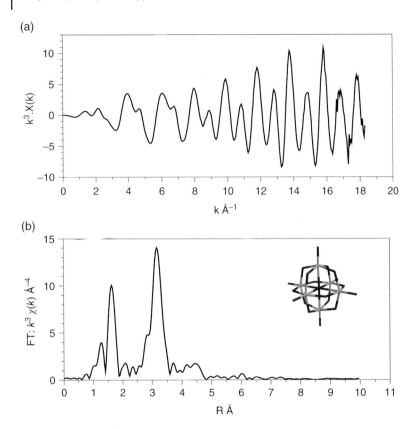

Figure 2.15 a) The $k^3\chi$(k) of the W L_3 edge of (NBu$_4$)$_2$[W$_6$O$_{19}$] (inset) and b) the magnitude of the Fourier transform of this pattern (*Source:* Diamond, B18, data from Richard Ilsley).

$$\Gamma\tau \cong \hbar = 10^{-16} \, eV.s \tag{2.8}$$

2.2.1   X-Ray Fluorescence

As an example, the *1s* core hole created above the *K* absorption edge of an element can be filled with an electron dropping down from a higher core level. A dipole-allowed transition, from a *2p* or *3p* dropping down to the 1s core-hole can result in the emission of x-ray fluorescence (Figure 2.16). Both of these pathways occur and give rise to different x-ray emissions with energies characteristic of the element. Different nomenclatures of x-ray emissions exist. The labels start with the core hole to be filled. So the prominent emissions for copper that are emanate from anode sources in many x-ray diffractometers, *Kα$_1$* and *Kα$_2$*, are also labeled by the destination and source states, *KL$_3$* and *KL$_2$*, respectively (Table 2.2).[2]

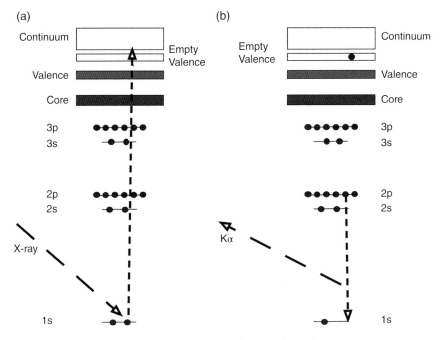

Figure 2.16 a) Creation of a core hole and b) its relaxation through x-ray emission.

Table 2.2 X-ray absorption edge energies and emission energies for copper.

Edge or emission	K	Kα₁	Kα₂	L₃	L₂
Transition label	K	KL_3	KL_2	L_3	L_2
Orbitals involved	$1s$	$1s\text{-}2p_{3/2}$	$1s\text{-}2p_{1/2}$	$2p_{3/2}$	$2p_{1/2}$
Observed energies (eV)	8980.5	8047.8	8027.8	932.5	952.5
Estimated from absorption edges (eV)	–	8048	8028	–	–

The close agreement between the measured x-ray emissions and those estimated from the absorption edge energies of the assigned transitions shows the validity of those descriptions.

X-ray fluorescence provides an alternative method to transmission to measure XAFS spectra (Section 4.3.3), and this can be the method of choice for dilute or thin samples. Under those conditions, the fluorescence yield is linearly related to the x-ray absorption. One of the key factors governing the sensitivity of the fluorescence yield sensitivity is the proportion of the core-hole that is relaxed through fluorescence rather than electron emission. For the K edges, this increases with a sigmoid curve from very small values for light elements (~2% for carbon) to over 90% for high Z elements (95% for gold) with the crossover at $Z = 30$ (Zn). Emissions related to the L absorption edges also increase in probability with atomic number.

2.2.2 Electron Emission

The alternative method accommodating the energy released through the filling of a core-hole by a higher energy electron is by electron emission (Figure 2.17). In this mechanism, the core-hole is filled leaving a vacancy in a higher core level. The electron from a third core level is emitted, and this is referred to as an Auger process, which leaves two core-holes. So, for example, at a K absorption edge, the K shell core-hole created ($1s$ orbital) could be filled by an electron falling from the L_1 shell ($2s$) and much of the energy released by the photo-ejection of a $2p$ electron is to overcome the binding energy, with the remainder being kinetic energy. The unpaired electrons in the resulting core levels can couple to give particular Russell Saunders states. In this example a transition would be labeled $KL_1L_{23}(^1P)$. The ion would have $2s$ and $2p$ vacancies, which couple to form a singlet state. Auger transitions can also involve valence orbitals (generally labeled as V).

There are many other secondary processes that cause electron loss. The current from all of these processes can be used to monitor x-ray absorption by

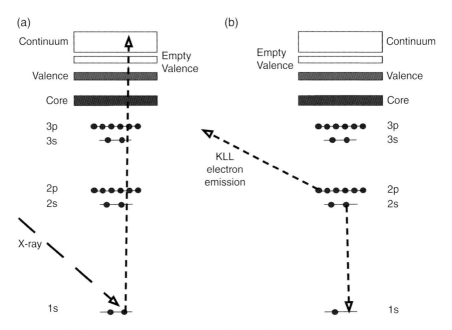

Figure 2.17 a) Creation of a core hole and b) its relaxation through Auger electron emission.

electron yield. In this simplest case the total current is monitored as a drain current from the sample. This can include a significant background, but sampling can be refined with energy selection right down to a single Auger transition of the x-ray absorbing element.

2.2.3 Resonant Inelastic X-Ray Scattering or Spectroscopy (RIXS)

Resonant inelastic scattering is also known as Resonant x-ray Raman. This process is shown in Figure 2.18. The incoming photon is of the appropriate energy to excite into the XANES region, either a low-lying valence or continuum state. The outgoing fluorescent photon may result from relaxation from the valence electrons into the vacated core hole or by a dipole allowed core to core-hole transition emitting a lower energy photon. The energy difference between incoming and outgoing photon energies is the inelastically transferred energy. So either the higher level core spectrum or even the valence band of a material can be probed with the energy source being that of the incoming x-ray, similar to vibrational spectra being probed by inelastic scattering of visible light in Raman spectroscopy.

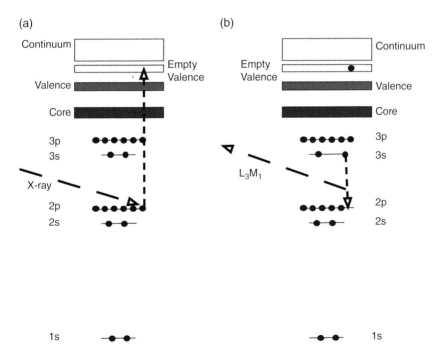

Figure 2.18 A x-ray resonant inelastic scattering process or Resonant x-ray Raman. a) L_3 edge absorption and b) L_3M_1 emission.

2.3 Effects of Polarization

With laboratory x-ray sources, the light has no polarization properties. Light from a synchrotron source is intrinsically polarized in the horizontal plane. Depending upon the viewing point, or the particular magnetic array type providing the light, polarization can be plane or circular.

2.3.1 Plane (Linear) Polarization

Many materials are isotropic in nature, being presented either as a solution or as a randomly oriented powder. For these cases, and for cubic crystals, the polarization in the light source is completely nullified. However, most crystalline samples or a surface will engender some anisotropy and the orientation of the polarization relative to the sample will become important. Then the orientation of a virtual state in valence band may cause the absorption to be reduced or lost. As in some NMR experiments, the isotropic spectrum can be reclaimed by magic-angle spinning with the sample oriented at ~54.7° (of cosine $1/\sqrt{3}$). In principle thorough polarization-dependence studies can provide orientational information, for example, of how molecules are adsorbed on a surface.

2.3.2 Circular Polarization

Circularly polarized x-radiation can also be sampled from storage ring light sources. It may be thought that this could be utilized to study optically active (chiral) materials in solution. However, there is no such effect observable in isotropic media. In crystalline samples, or in solution oriented in liquid crystal solvents, pre-edge features have been shown to display circular dichroism.

2.3.3 Magnetic Dichroism

The more prevalent application of circular polarized x-rays in XAFS spectroscopy is to investigate x-ray magnetic circular dichroism (XMCD). In ferromagnetic materials, it has been shown that the absorption at the L_2 and L_3 edges display XMCD effects in magnetic fields. Antiferromagnetic, ferromagnetic and ferromagnetic materials display magnetic dichroism with linear polarization (XMLD). Magnetic dichroism arises from coupling between the between the $2p^5$ electrons with the valence electrons in the excited states ($3d^{n+1}$). For example, for Ni(II) ($3d^8$), the excited state would be $2p^5 3d^9$. Right- and left-hand polarized light cause changes of +1 and -1 to the magnetic J quantum number, and it is by virtue of the J states that the L_3 and L_2 edges differ. An example of this effect is shown in Figure 2.19, recorded in total electron yield from a thin film sample.

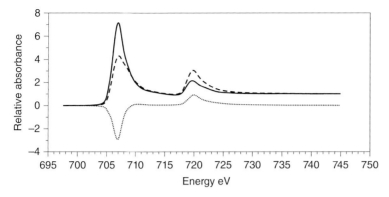

Figure 2.19 Example of the XMCD effect (dotted) at the L_3 and L_2 edges of an iron oxide. XAFS with right (solid) and left (dashed) polarized light. Recorded by total electron yield on a thin film sample (*Source:* Diamond, I06, data from Sarnjeet Dhesi).

2.4   Questions

1   The energies and linewidths of absorption edges of molybdenum are given in Table 2.3.

   A Estimate the energies of the following emission lines: $K\alpha_1$, $K\alpha_2$, $K\beta_1$, $K\beta_2$.

   B From the data in the table, assess which edges may provide most information in the XANES and EXAFS region. What would be the observable k range for the L_3 edge (see equation 2.4)?

2   The K absorption edge of sulfur occurs at 2472 eV and the M_1 to M_5 edges of lead at 3850.7, 3554.2, 3066.4, 2585.6, and 2484.0, respectively. For the compound $PbMoS_4$, which edge(s) could be studied without complication from the other elements, and over what energy range? Which edge might be the best to pick to provide some EXAFS data to $k = 8\,\text{Å}^{-1}$?

Table 2.3 The energies and line widths (eV) of the *K, L,* and *M* absorption edges of molybdenum.

Edge	*K*	L_1	L_2	L_3	M_1	M_2	M_3	M_4	M_5
Energy	20000.4	2865.5	2625.1	2520.2	506.3	411.6	394.0	231.1	227.9
Width	4.5	4.25	1.97	1.78	~6	2.2	2.2	0.16	0.14

3 The structure of Na_2MoO_4 contains tetrahedral MoO_4^{2-} anions, with a Mo-O distance of 1.97 Å. Using equation 2.6, how many complete oscillations should be observed if a k range of $12\,Å^{-1}$ were recorded?

4 The related salt $Na_2Mo_2O_7$ has a chain structure with two Mo sites in the crystal: one tetrahedral and the other octahedral.
   A How might these be distinguished in the XANES features of the K and L_3 edges?
   B The Mo-O distances in the four-coordinate tetrahedral site have an average value of 1.756 Å with a mean square deviation of $0.004\,Å^2$. The six-coordinate octahedral site has a larger average (1.95 Å) and spread ($0.227\,Å^2$).
   C Use equation 2.7 for these two sites, taking the static disorder as σ^2 and plot χ up to $k = 15\,Å^{-1}$.
      Note the difference in the observable k range in these two cases.
   D The octahedral site is made up of three pairs of Mo-O bond lengths of 1.685, 1.899, and 2.267 Å, which may be considered as three different shells. Remodel the χ function as the sum of these three shells each with a coordination number of 2, taking $2\sigma^2$ as $0.005\,Å^2$. Note the beats between the shells and that the simple one-shell model of the octahedral site misses out much detail.

References

1 'Tables of x-ray mass attenuation coefficients and mass energy-absorption coefficients from 1 keV to 20 MeV for elements $Z = 1$ to 92..', J. H. Hubbell, S. M. Selzer, www.nist.gov/pml/data/xraycoef/index.cfm.
2 'X-ray Wavelengths', J. A. Bearden, *Rev. Modern Phys.*, 1967, **39**, 78–124.
3 'X-ray Data Booklet', Ed. A. C. Thompson, Center for X-ray Optics and Advanced Light Source, LNBL/PUB-490 Rev. 3, 2009. (http://adb.lbl.gov/xdb-new.pdf). On-line calculations from http://henke.lbl.gov/optical_constants.
4 'Cr K-edge XANES: ligand and oxidation state dependence. What is oxidation state?', M Tromp, J. Moulin, G. Reid, J. Evans, *AIP Conf. Proc.*, 2007, **282**, 699–701.

3

X-Ray Sources and Beamlines

3.1 Storage Rings

As described in Section 1.5, the applicability of XAFS was transformed by the availability of synchrotron light sources in the 1970s. This spectroscopy, like other types of experiments carried out in parallel, including diffraction, was performed in parasitic mode. In other words, the synchrotron light was exploited as a side benefit, but these experiments did not have priority in terms of the operational parameters of the machine. Hence, availability, beam lifetime and stability could be compromised. Facilities used in this phase were termed "first-generation sources."

3.1.1 Second- and Third-Generation Sources

The scientific advances made in the first-generation era were sufficient to produce a demand for dedicated light sources, termed the "second generation of light sources." These sources developed through the 1980s, and they consisted of a ring of bending magnets to deflect charged particles, normally electrons (positrons are used in some storage rings), along their path around the ring. Their characteristic was to provide light sources from the bending magnets that formed the circle of the particle storage ring. Early in their lifetime, the need to change and enhance the light source available to be appropriate for particular experiments was recognized. In response, new magnetic arrays, called insertion devices, were added into the straight sections linking the bending magnets. This approach was so successful that by the 1990s, "third-generation" light sources were being built and commissioned. A schematic of a synchrotron source is presented in Figure 3.1a, with one realization, Diamond (Oxfordshire, UK) shown in Figure 3.1b. Their characteristic was that the principal light sources came from insertion devices, and the primary purpose of the bending magnets was to complete the ring. However these magnets too also provide very effective light sources, both in the x-ray and longer wavelength ranges (from the

X-Ray Absorption Spectroscopy for the Chemical and Materials Sciences, First Edition. John Evans.
© 2018 John Wiley & Sons Ltd. Published 2018 by John Wiley & Sons Ltd.

(a)

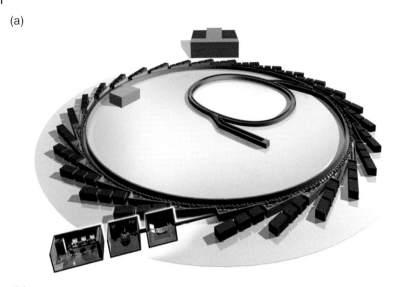

(b)

Figure 3.1 a). Three-dimensional model of Diamond showing the positions of the linear accelerator, the booster synchrotron, storage ring, beamlines, hutches. and control cabin. b). An aerial view of the Diamond site that can be correlated with the schematic source (*Source:* Courtesy of Diamond).

vacuum ultraviolet to the infrared). National and international centers have grown up across the world. A central repository for information about these facilities is provided by lightsources.org.[1] This includes links to the properties and application modes of the available light sources.

3.1.2 Bending Magnet Radiation

The nature of the magnetic array and the electron energy of a storage ring define the headline characteristics of a light facility. The light from a bending magnet has an extremely broad energy spread. This results from the very short viewing time of the light from electrons as they pass the viewing point at virtually the speed of light. The critical energy, E_c (keV), of a storage ring depends upon the magnetic field of the bending magnet, B (T) and the electron energy, E_e (GeV). In practical units, this relationship is as in equation 3.1:

$$E_c = 0.6650 E_e^2 B \qquad (3.1)$$

Bending magnet fields are generally in the range of 1 ± 0.5 Tesla so the main influence on the energy profile is the electron energy. This is shown in Figure 3.2, comparing two of the third-generation sources in the United States:

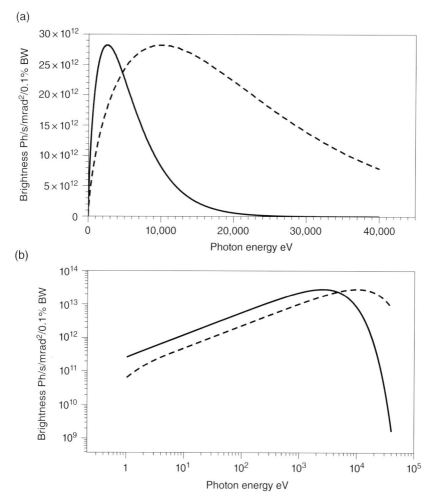

Figure 3.2 Plots of the brightness output from a bending magnet. Plots a) and b) are per the Advanced Light Source (ALS): E_e 1.9 GeV, Current 400 mA, B 1.27 T and c) and d) as per the Advanced Photon Source (APS): E_e 7.0 GeV, Current 100 mA, B 0.6 T. The dashed lines are for a 5 T magnet at the ALS.

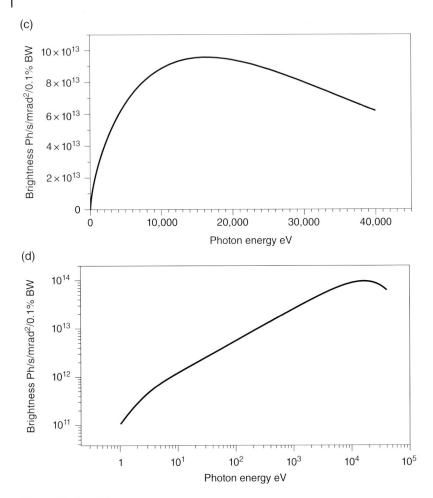

Figure 3.2 (Cont'd)

the ALS (Berkeley, E_e 1.9 GeV) and APS (Argonne, E_e 7.0 GeV), calculated by equation 3.1 to have E_c values of 3.0 and 19.6 keV, respectively. Such plots are readily generated online using reference.[2] The critical energy is the value where 50% of the power is of lower and higher photon energy, and is close to, but slightly higher in energy than, the energy of maximum output. The output is shown in units of photon/s/mrad^2/0.1% bandwidth, generally referred to as *brightness*. The linear plots show the energy ranges of maximum output clearly. These range over several keV for the ALS, and several tens of keV for the APS. It is more conventional though to see these data as logarithmic plots. This better shows the very wide extent of the available spectrum range of storage ring sources, but makes the energy region of maximum performance less clear.

Clearly, the ALS has a lower energy (softer) output spectrum than the APS with its much higher electron energy. Storage ring architecture is based upon a number of cells. The simplest system would have a single bending magnet is each cell. Double and triple bends are more common designs in third-generation sources, and yet more complex patterns are now being installed. The triple bend eases the way for making the central bending magnet of substantially different magnetic field, without perturbing the electron path at the end of the cell. The field of the central bending magnet can be increased to move the output spectrum to increase the flux of hard (high energy) x-rays. The result of a superconducting 5 T "superbend" magnet at the ALS, which raises E_c to 12 keV, is shown in Figure 3.2; this approach is also adopted for the SuperXAS beamline at the Swiss Light Source (SLS) (E_e 2.4 GeV, B 2.9 T, E_c 11.1 keV). It shows one way in which the source at the storage ring can be modified to better match the science at a particular beamline. In this case there is a significant gain in available light at increased photon energies, at the expense of having less light at lower energies.

Another figure of merit for storage ring light sources is the *emittance*, ε, which is the product of the electron/positron beam size and its divergence in the horizontal plane (σ_x and $\sigma_{x'}$, respectively) (equation 3.2). The emittance in the vertical plane is reduced by a coupling factor, κ, typically about 1%, but may approach 0.1%. Thus lower emittance values will provide the basis of a more focused photon beam. For an early second-generation source the emittance values were over 100 nm.rad, reducing to <5 nm.rad for current third-generation sources (e.g., 2.84 nm.rad for Diamond), and to about 1 nm.rad for new machines (e.g., PETRA III (Hamburg) and MAX IV (Lund)).

$$\varepsilon_x = \sigma_x . \sigma_{x'} \tag{3.2}$$

The output of light sources is generally presented (brilliance as defined is termed brightness in some reports) in terms of:

Flux: photons/s/mrad/0.1% bandwidth
Brightness: flux/mrad
Brilliance: brightness/mm^2
Power: W/mrad

For an experiment the important factor is the flux on the sample, and the importance of these different output parameters varies with the size of the sample. For large samples, the flux gives a good guide about the available light, and the depends upon the particle current I_b, the electron energy, E_e^2, the opening angle of the photon beam and the "universal curve" of output of the shapes shown in Figure 3.2. The power, P, is more problematical (equation 3.3), and largely represents heat onto the beamline optical elements.

$$P = 4.22.10^{-3} E_e^3 I_b B \quad Wmrad^{-1} \tag{3.3}$$

For very small samples, the brilliance parameter is the key, representing as it does the potential flux density on that sample.

A benefit for XAFS studies of storage ring output is the broad, smooth variation in output. The observed spectral range, including pre- and post-edge regions, might extend up to 2000 eV, and, as we have seen, EXAFS oscillations at high k values become very weak. Hence, having an intense source with a smoothly curved background is a good option for collecting optimum EXAFS data.

3.1.3 Insertion Devices

Insertion devices (IDs) are more complicated magnetic arrays that are located in the straight sections of storage rings. They deviate the path of the electrons (or positrons) off the axis of the central path and create oscillations in the particle path. Light is again emitted from the tangent points and follows a path straight from those points. A downstream bending magnet curves the electron path out of the light path to allow the x-ray to be extracted into a beam line. The purpose of their installation is to provide a better light source for the type of experiments located on the corresponding beam line.

3.1.3.1 Wavelength Shifters and Multipole Wigglers

The simplest of the devices is a *wave-length shifter* or *wiggler*. It generally involves three poles of a superconducting magnet, reducing the radius of curvature of the electron path, and thus increasing the critical energy, E_c, of the output spectrum. The deviation in the electron beam path is parameterized by the *deflection factor, K* (equation 3.4). The degree of deflection depends upon magnetic field, B (Tesla), and the wavelength of the magnetic array causing the oscillation in the path, λ_u (mm). For a wavelength shifter, K is relatively large (>3). The effect is essentially the same as a super-bend, and the critical energy can be calculated in the same way (equation 3.1). As with a super-bend, there is an increase in flux at higher photon energies and a decrease at longer wavelengths.

$$K = 0.09337 B\lambda_u \tag{3.4}$$

If a series of oscillations is set up in a *multipole wiggler*, then there are independent light sources from each of the tangent points as the emission points are outside the photon beam from the upstream magnets. The extra source points increase the total flux and brilliance in proportion to the number of poles, as well as just moving the spectrum. The output power similarly increases and the smooth variation of output with energy is still retained when $K \gg 1$.

3.1.3.2 Planar Undulators

If the deflection factor is reduced, generally by reducing the magnetic field, B, then the electron oscillations stay within the cone of radiation generated at the start of the undulator. The sources then become coupled and the result is a series of harmonics of fundamental energy E_n (keV) in the output, which, when viewed on-axis, have an energy dependent upon the electron energy, E_e, and the properties of the undulator: the periodicity and magnetic field at the particle beam (equation 3.5).

$$E_n = \frac{9.496nE_e^2}{\lambda_u \left(1 + \frac{K^2}{2}\right)}$$

(3.5)

An example of the output of an undulator is shown in Figure 3.3. The width of the harmonics may accommodate XANES scans but are not ideal for EXAFS

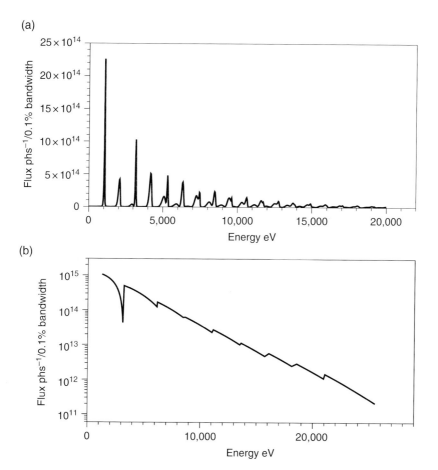

(a)

(b)

Figure 3.3 The output of the 27-mm period undulator (2 m) on I18 of Diamond operating at 300 mA. a) The harmonics from a 7 mm gap. b) The output curves derived with optimized gap scans.

since the flux on a sample will vary greatly across the spectrum; anyway its output will probably not correspond to the energies of absorption edges of interest. Undulators generally consist of arrays of permanent magnets, so changing the electron energy, the periodicity of the magnets, or their intrinsic magnetic field are not options for tuning the output of the insertion device as a spectrum is being set up or scanned. However, this can be achieved by scanning the physical gap between the undulator magnets and the electron beam, thus changing the magnetic field at the electron beam. By increasing the gap, the effective field will be reduced, reducing the deflection factor K. This will *increase* the energy of the fundamental and its harmonics, albeit at the expense of reducing the output intensity somewhat. In practice undulator gaps can be controlled to about 1 μm and gap scanning, which couples the motions of the monochromator and the undulator, does provide a way to record EXAFS.

The major effect of the undulator array is to increase the brilliance, by a factor of N^2 in the ideal case. The result is an output spectrum as exemplified for a 2-m-long undulator with a 27-mm period installed at Diamond on I18 (Figure 3.3). At a K value of 2, the fundamental is calculated to have an energy of 1.05 keV. The odd harmonics are the more intense and they decay in intensity with increasing harmonic number. By altering the gap an optimized performance for the light output can be derived across a wide spectral range (1–20 keV) without any breaks in the spectrum.

This light source forms the first element of a microfocus beamline, which is shown in Figure 3.4. This 27-mm undulator has a nominal gap between the poles of 7 mm. In such a case the space for the walls of the vacuum chamber

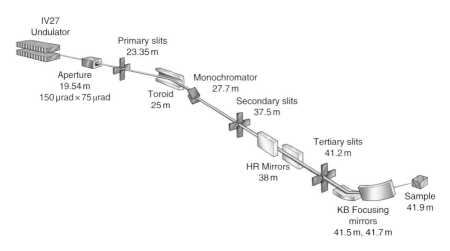

Figure 3.4 Schematic of the components of the Microfocus Spectroscopy beamline, I18, at Diamond with a 27-mm period in vacuum undulator as source (*Source:* Courtesy of Diamond).

containing the electron beam is very restricted so the whole undulator magnetic array is housed within the vacuum chamber and the gap is changed by drive shafts passing through seals into the storage ring vacuum chamber. Accordingly such an undulator is described as an IV, in vacuum, device. The difference with out-of-vacuum devices is generally the IV undulators have higher photon energy outputs coming from shorter periodicities, and smaller gaps are required to achieve the deflection with smaller pole pieces.

3.1.3.3 Helical Undulators

In all of the devices described so far, from bending magnets to planar undulators, the charged particle motion has been in the machine plane, that is, horizontal. As a result, the emitted radiation in that plane is polarized in that plane to a very high extent. Above and below the plane, there is a component that is circularly polarized and dichroism experiments can be carried out by taking the difference between spectra obtained when sampling with light from above and below the machine plane. However, a higher degree of polarization control can be achieved by a more complicated magnetic array, such as a helical undulator. These have four arrays of magnets that can be moved independently, as for beamline I06 at Diamond (Figure 3.5). This beamline is designed to provide x-radiation in the soft x-ray region (80–2100 eV), requiring a longer period undulator (64 mm) and allowing a larger gap from the permanent magnets of the undulator (15 mm). Hence this undulator structure can be outside the vacuum chamber of the storage ring. The phase of the magnetic fields felt by the electron beam can be altered through the moving the relative positions of the four arrays of magnets along the beam direction. Thus the electron motion can be changed to an oscillation including motion in the vertical plane; vertically polarized light is invaluable for surface studies on solids and liquids. In addition to planar polarization, circular polarization of either sense can be created. In this way, the x-ray magnetic circular dichroism (XMCD) effects shown in Figure 2.16 can be achieved.

3.1.4 Time Structure

The electron current is made up of a set of bunches of electrons circulating around the storage ring at close to the speed of light, so the circulation time depends upon the circumference of the storage ring. For a large ring like the ESRF (Grenoble) (844.4 m), the transit time is 2.816 µs, with each bunch of length of about 100 ps. But at the maximum operating current of 200 mA many bunches are injected (up to 992 at the ESRF), and the gap between bunches can be reduced to 2.82 ns.

The majority of XAFS measurements are made over timescales of seconds or minutes. As a result, this pulsing is not noticed, and the source is effectively continuous in time. However, this time structure can be exploited, and storage

(a)

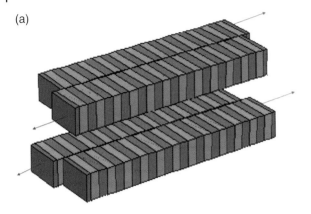

(b)

Figure 3.5 A helical undulator, as on beamline I06 at Diamond (*Source:* Courtesy of Diamond) a) schematic of the magnetic array, b) the magnetic array of an undulator assembly.

rings are operated with a variety of hybrid bunch-structures in which a section of the filling pattern has a multibunch injection and a portion contains a single, larger bunch. This allows the potential for simultaneous acquisition in a quasi-continuous and a pulsed mode at different beamlines.

For the pulsed experiments, the storage ring provides a master clock and experiments can be timed against this to utilize the isolated bunch. The normal time resolution can be then the time for the bunch to traverse the viewing point, typically 10s ps to 100 ps. This is much longer than the timescale of molecular motions during chemical reactions, and thus is not fast enough to create a real-time "movie" of the reaction. It can, however, identify short-lived transients, and probe the *kinetics* of their transformations, a measure of the probability of the reaction occurring under those conditions. This and other time-resolved options have been discussed in a review.[3] Any further reduction in the observation time on a storage ring is at a cost in photon flux on the sample. The bunch can reduced in time by reducing the number of electrons in it and by operating the storage ring in a particular way (low alpha mode), by rotating the electron bunch through a section of the storage ring (called a crab cavity), by using a streak camera as a detector to "slice" into the output from the sample, or by actually take a slice out of the bunch by deflecting the charged particles with an intense laser.[4] It is the last of these that has achieved the best time resolution (<250 fs),[5] but the others provide options up to the 5 ps range.

3.2 Other Sources

3.2.1 Laboratory Sources

A laboratory anode x-ray source does provide a broad x-ray background spectrum in addition to the x-ray fluorescence lines used for x-ray diffraction. This broad background, called *brehmsstrahlung*, is about 10–20% of the intensity of the fluorescence lines at the energy of its maximum output. It is caused by the electron hitting the anode target and losing some of its energy ("braking"). The maximum energy emitted will be the energy of the bombarding electrons, but the maximum output is at 50% of that value. The proportion, P, of the total electron energy that gives rise to the *brehmsstrahlung* emission has been found to be related to the atomic number of the target element, Z, and the accelerating voltage, V (volts), as in equation 3.6.

$$P = 1.1.10^{-9} ZV \qquad (3.6)$$

For XAFS purposes, the best target would be a high Z element, for example, tungsten, with a high accelerating voltage and a high current. However, the brilliances are orders of magnitude less than achieved using a storage ring source. Such sources continue to improve in terms of brightness and now begin to approach the brightness of second-generation sources.

In one laboratory device, a very compact electron storage ring[6] (radius 0.16 m), designed as an infrared source, has positioned within it a micrometer sized metal target, for example, Cu or W. The high-energy electron beam generated *brehmsstrahlung* with significant intensity from a few keV to a few MeV and XAFS measurements can be made on such a source.

3.2.2 Plasma Sources

A metal or a ceramic target can also bombarded by an intense laser source to create a plasma with highly excited ions that relax by x-ray emissions. The size of the emission source has the characteristics of the laser spot, and the time structure also follows that of the laser. Early experiments[7] utilized nanosecond pulses to record XAFS spectra of a metal (aluminum), but the pulses can be considerably shorter than the timescale than a storage ring pulse. Shot-to-shot reproducibility of the source becomes important in the applicability of the technique for XAFS studies of more complex materials. X-ray sources can also be created by the plasma formed by pulsed electron beams. These sources can be dominated by x-ray emission lines.[2] Very high plasma temperatures are required to achieve significant *brehmsstrahlung*.

3.2.3 High Harmonic Generation

An alternative approach to ultra-short x-ray pulses (~20 fs) is to utilize a femto-second laser as the fundamental and generate very high harmonic numbers (several 10s).[8] The purpose is to interrogate the structure of a material faster than the period of a molecular vibration. Then the *dynamics* of a structure may be monitored as atomic motion is taking place, giving a "snap-shot" of a molecular motion or chemical reaction. This involves passing the light at 800 nm wavelength from, for example, a Ti:sapphire laser through a capillary containing a gas, generally an inert gas. Fluxes similar to storage ring sources (10^{12} photons/s) can be generated in the extreme ultra violet energy range (~50 eV) with femtosecond time resolution. These sources are fully coherent. There are medium term aims to extend the energy range up to 1 keV, which would include the *K*-edges of C, N, O, and F, and the *L* edges of the *3d* transition metals.

3.2.4 Free Electron Lasers (FELs)

Ultra-short x-ray pulses have now been generated in Germany, the United States, and in Japan by Free Electron Lasers (FELs), which are an extension of undulator technology. The FLASH source in Hamburg has been operating since

2005 from ~28 eV to 300 eV, with pulses of 10–100 fs. The light delivered by a FEL in one pulse is similar to that from a storage ring in 1 second, thus opening up structural studies in the dynamics time regime to a broader scope of materials. In April 2009, the first x-ray FEL pulses came from the LCLS at Stanford in California at 1.5 Å wavelength (8275 eV) and in June 2011, x-ray lasing at 1.2 Å (10300 eV) was achieved at the SACLA FEL in Japan. Other sources are in train, with the European XFEL (Hamburg), due to open for user operation in 2017, designed to produce x-radiation at energies up to ~25 keV with less than 100 fs pulse length. The first hard x-ray beamline (1.7–12.3 keV) of Swiss FEL facility, SwissFEL is also planned to commence operation in 2017. The peak and average brilliances planned for the XFEL of 10^{33} and 10^{25} photons/s/mm^2/mrad^2/0.1% BW, respectively, exemplify how will transform the possibility for x-ray studies of atomic motion. However, the output of undulators is not readily matched to the broad energy range of EXAFS measurements. Rather XANES and x-ray emission spectroscopy techniques (e.g., RIXS) can be expected to be important techniques early in x-ray FEL applications.

3.3 Beamline Architecture

Although there are other options, as outlined in the previous section, storage ring sources will be the location of the majority of XAFS measurements and experiments for many years ahead. This section will describe key elements of the optics of a beamline to provide the light on the sample. The options for detectors will be discussed in Chapter 4. Beamline architecture has developed markedly since the early spectrometers of first- and second-generation sources. That simple outline is presented in Figure 3.6. After the source, one would find a set of slits to define the beam before the crystal monochromator, often used as the sole element to provide monochromatic radiation following Bragg's law (equation 3.7). Following the monochromator there are detectors to measure the x-ray flux entering the sample, leaving the sample and then one leaving a reference material. $I(o)$ and $I(t)$ are used to define the XAS of the sample, and

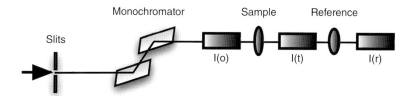

Figure 3.6 Schematic for an early XAFS beamline. $I(o)$, $I(t)$, and $I(r)$ detect the x-ray flux before the sample, after the sample and after the reference material.

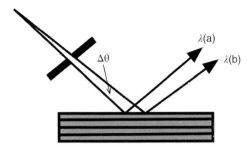

Figure 3.7 Angular divergence of a beamline with a source to slit distance of L and a vertical slit width of s_v and a source size of σ_v.

$I(t)$ and $I(r)$ that of the reference material. The reference is used to provide an internal energy calibration for the spectrum.

$$n\lambda = 2dsin(\theta) \tag{3.7}$$

The emittance values of earlier sources caused the photon source to be larger and more divergent than for third-generation light sources. The important factors contributing to the resolution provided by the spectrometer were the source to pre-monochromator slit distance and the vertical slit aperture. In this simplest of optical arrangement, the size of the source in the vertical plane (σ_v), the distance from the source to slits (L), and the size of the vertical slits (s_v) will define the divergence (Figure 3.7) on the monochromator crystal (equation 3.8), and the energy resolution ($\Delta E/E$) resulting from the beam divergence and the intrinsic width of the reflection ($\Delta\theta_B$) is given in equation 3.9, as shown in Figure 3.7.

$$\Delta\theta = \sqrt{\left\{\left(\frac{\sigma_v}{L}\right)^2 + \left(\frac{s_v}{L}\right)^2\right\}} \tag{3.8}$$

$$\frac{\Delta E}{E} = cot\theta \sqrt{\left(\Delta\theta^2 + \Delta\theta_B^2\right)} \tag{3.9}$$

Figure 3.8 shows the contribution to the energy resolution a Si silicon monochromator with a beam size of a second-generation source and pre-monochromator slits set 18 m from the source point. Over most of the energy range, the source size is a dominant factor. It can be seen that the resolution is improved by ⅓ by reducing the slit aperture from 1 mm to 0.1 mm. A further reduction of a factor of 2 is possible by changing the monochromator crystal to the higher (Miller) index reflection. For each crystal plane, then the intrinsic resolution from the reflection off the crystal (the Darwin width) is much greater and contributed relatively little to the overall resolution (Figure 3.8b). This figure shows the source size influences the choice of slit size and monochromator crystal significantly, especially if wishing measure XANES features with the natural linewidth of the spectral features. In this configuration only the Si(311) monochromator with 0.1 mm slits could approach the core hole

(a)

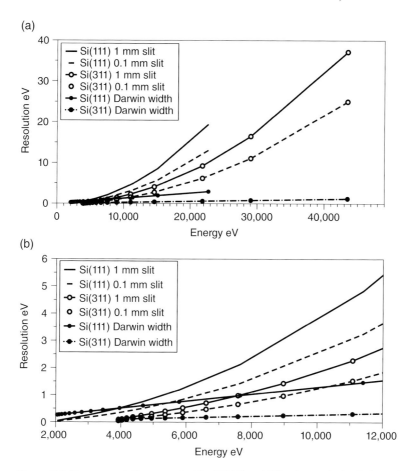

(b)

Figure 3.8 Energy resolution expected for Si(111) and Si(311) crystals divided into the Darwin width [9] contribution and from beam divergence from a 900 μm vertical source size and pre-monochromator slits of 1 and 0.1 mm set at 18 m from the *Source:* a) energy range available between 5 and 75° and b) for 2 to 12 keV.

lifetime linewidths of the Zn (1.7 eV at 9.66 keV) and Mo (4.5 eV at 20.0 keV) *K*-edges. The situation for EXAFS measurements with broader, weaker features is different. The flux delivered by the Si(311) monochromator is only ~10% of that from the Si(111) meaning that a practical compromise is generally used.

The characteristics of a XAFS beamline on a third-generation source are rather different. The vertical photon source size in the vertical plane from a bending magnet source is much smaller; for example, 16 μm on the SuperXAS beamline at the SLS, and beamline design is more complex. The main optical elements can be followed on a beamline layout example shown in Figure 3.4.

Following the aperture to extract the central cone of the undulator, there is a set of slits that define the light passing through the first element, an x-ray mirror.

3.3.1 Mirrors

X-ray mirrors are generally based upon a silicon structure and mostly have metal coatings with a very high degree of flatness. Their purposes are to collimate and focus the x-ray beam and also to provide harmonic rejection. In this example, the first mirror has a toroidal shape, designed to collimate in the vertical plane and to focus the x-rays in the horizontal plane at the secondary slits, which become a virtual source point. The collimation in the vertical plane gives a more parallel beam to enter the monochromator and thus optimize the energy resolution attainable. The coating of this mirror is rhodium. The K absorption edge of this metal (23.2 keV) is higher than the energy range of the beamline, and thus will not perturb the light reflected by the mirror. Because of the absorption of energy, the mirrors are generally cooled to maintain stable properties. A "heat-bump" can provide a dispersion of incident angles that will degrade the resolution.

The second pair of mirrors along the light path is for harmonic rejection. According to Bragg's law (equation 3.7), not only the fundamental but also harmonics ($n > 1$) can pass through a monochromator. An undulator source will naturally create higher harmonics, and both bending magnet and wiggler radiation may also have significant intensity at twice and three times the desired photon energy. These higher harmonics will distort the XAFS spectrum; since they will be more penetrating through the sample, they do not contribute to the absorption edge and they have a different sensitivity at the detectors. Harmonic rejection mirrors are set so that reflectivity of the first harmonic ($n = 1$) is high, but the more penetrating radiation of the higher harmonics are not externally reflected and thus absorbed by the mirror. The reflectivity can be controlled by the atomic number of the coating (higher Z elements have higher critical angles) and the angle of the grazing incidence. Example curves are given in Figure 3.9 for the two coatings laid down on the harmonic rejection mirrors in Figure 3.4, namely nickel and rhodium. Often two coatings are chosen so that the spectral range of the beamline can be covered by a harmonically pure x-ray beam, without intensity distortions near an absorption edge of the mirror coatings,

From Figure 3.9, it can be seen that:

1) Changing from Ni to Rh increases the energy at which high reflectivity can be achieved,
2) Decreasing the grazing incident angle also increases the critical energy.

The effects of the Ni K edge (8.3 keV) and Rh L edges (near 3 keV) are very evident. Taking the example the scandium K edge (4492 eV), it can be seen that

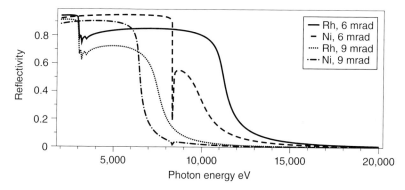

Figure 3.9 Reflectivity of nickel and rhodium mirrors set at two angles. Surface roughness set at 1.5 Å RMS.

a setting of a nickel mirror at 6 mrad (0.344°) could give good reflectivity for an EXAFS scan of 1000 eV above the absorption edge and also achieve harmonic rejection. It also demonstrates that one mirror setting will probably serve well for selected range of absorption edges, but must be modified for other studies.

3.3.2 Monochromators

Monochromators operate on the principle of the Bragg equation (equation 3.7). From 1000 eV (λ = 12.4 Å) upward in photon energy, a single crystal can provide d spacings that allow the scanning of XAFS spectra by altering the Bragg angle, θ. For photon energies up to 2000 eV (λ = 6.2 Å), gratings can be used, as they have larger spacings (Table 3.1). Synthetic multilayer materials, consisting of alternative layers of materials of significantly different atomic number, for example, Mo/C/Si or WSi_2/Si, have also been employed, but

Table 3.1 Crystal planes used in XAFS spectrometers and the energy ranges accessible with a wide monochromator scan range.

Crystal Plane	2d (Å)	Energy range (keV) $5° < \theta < 75°$
Si(311)	3.27	3.9–43.5
Si(220)	3.84	3.4–37.1
Si(111)	6.27	2.1–22.7
Ge(111)	6.532	2.0–21.8
InSb (111)	7.48	1.7–19.0
YB_{66} (400)	11.75	1.1–12.1
Beryl (10 $\bar{1}$ 1)	15.95	0.8–8.9

generally they provide poorer energy resolution. These structures can also be incorporated into gratings.

$$E = \frac{12.3984}{2d\sin(\theta)} \quad keV \tag{3.10}$$

In terms of energy range, Bragg's equation can be reset into equation 3.10, with d in units of Å. This directly provides the energy range attainable using a given monochromator angle range, which is the first parameter necessary for choosing the right monochromator for a given experiment. Other key materials characteristics are crystal quality to optimize resolution and stability under operating conditions; radiation damage is generally the most significant problem. The photon energies provided by two common monochromator crystals, Si(111) and Si(311), at different Bragg angles are shown in Figure 3.10. These different cuts of silicon provide the different d spacings, which can be calculated from the different Miller indices (*hkl* values) of the crystal planes (equation 3.11 for a cubic system):

$$\frac{1}{d^2} = \frac{1}{a^2}\left(h^2 + k^2 + l^2\right) \tag{3.11}$$

This shows that higher indexed planes will afford smaller d spacings, matching to higher photon energies. The slightly larger unit cell of germanium ($a = 5.64613$ Å)

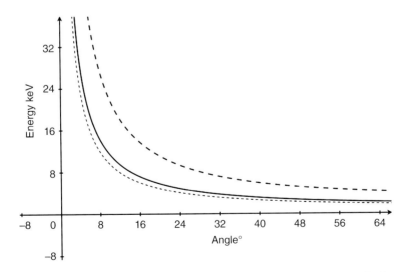

Figure 3.10 Photon energies attained by different Bragg angles for Si(111) (full line) and Si(311) (dashed) and a InSb(111) (shorter dashed).

Figure 3.11 Double crystal monochromator.

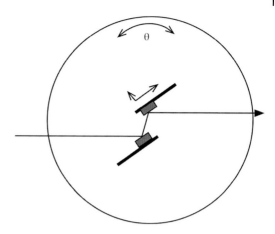

as compared to silicon (a = 5.43095 Å) accesses only slightly lower energy edges, and indium antimonide (a = 6.4794 Å) extends that range rather further.

As shown in Figure 3.4, the monochromator has two crystal faces with x-rays impinging on a reflecting (or Bragg) geometry. This can be either in the from of a single crystal with a gap cut into it for the beam to pass (*channel-cut crystal*) or as two separate crystals (*double-crystal*), which allows for more freedom of relative orientations. The coupling of the two crystals allows the monochromatic beam to continue along the beamline in a horizontal path with a vertical displacement (Figure 3.11). As the crystal is rotated during a XAFS scan, this vertical displacement would change as the incident angle is altered. In a *fixed-exit double crystal monochromator* there are additional relative motions of the two crystal faces to offset this effect and maintain the beam location on the sample.

An alternative monochromator arrangement is a transmission, or Laue, geometry (Figure 3.12). The example shown is of a double crystal monochromator with curved crystals on intersecting Rowland circles. The intersection acts as a virtual source point for the reflected beam from the first crystal. There are advantages of this approach over Bragg monochromators at high energy (> ~ 60 keV).[10] This is rarely used for XAFS spectrometers, which mostly operate at lower energies than that. There is a special case for a single bent Laue monchromator for energy dispersive configuration at energies > 20 keV, when this is beneficial for providing a small focal spot.

In a double crystal monochromator, the first crystal takes the load of removing the majority of the unwanted radiation exiting the first mirror. Thus it is subject to significant heating, which will change the d spacing of the crystal. This can affect the energy reflected and also broaden the energy spread of the reflected beam. If the temperatures of the two crystals differ then there will be a miss-match of their d spacings. Distortions of the monochromator may also move the beam position

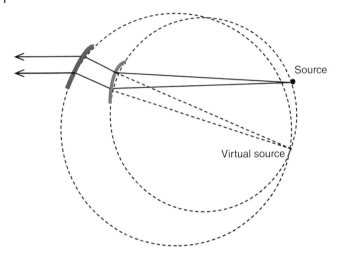

Figure 3.12 Geometric arrangement for a Laue monochromator.

on the sample. Thus the practice is to cool the crystal either by water or liquid nitrogen, depending upon the heat load to be addressed. A particular advantage with cryogenic cooling is that the thermal expansion coefficients of silicon cross zero at 120 K, and thus the d spacing is insensitive to small temperature changes around that temperature. At liquid nitrogen temperatures the thermal conductivity of silicon is also significantly higher (Conductivity, κ, = 150 and 1000 W/m/K at 300 and 80 K, respectively), an added advantage of cryogenic cooling.

The resolution of a well-cooled, perfect crystal monochromator will be determined by three main parameters:

1) monochromator angular step
2) x-ray beam divergence and
3) intrinsic (Darwin) width of the reflection.

As can be seen from Figure 3.10, the energy sensitivity for an angular movement varies considerably between low and high energies. An attainable angular resolution of 1 μrad ($5.7.10^{-5}$ ° or 0.21 arc sec) for a monochromator means that this is not a limiting factor. The energy steps at different Bragg angles are presented for three crystal planes in Table 3.2. Evidently, the step resolution improves with higher Bragg angles, but even at the shallowest of angles the energy resolution ($\Delta E/E$) is 1×10^{-5} or considerably less and the energy step is less than the core hole broadening of edges at these energies. Scan steps can be in multiples of millidegrees, and so these are also presented in the table. This shows that this is comparable to the core-hole lifetimes in that region of the spectrum and thus is too coarse a step for XANES measurements at low Bragg angles. In recent years, beamlines have been fitted with energy encoding

Table 3.2 Monochromator crystal planes showing the energy steps (eV) per millidegree of rotation at a series of Bragg angles (°), the Darwin width of the plane and the broadening due to core-hole lifetime of a nearby *K* edge (energy in keV).

Crystal plane	Bragg angle	Energy (keV)	Energy eV/mdeg	Darwin width (eV)	Lifetime effect (eV)
Si (311)	7	31.10	4.4	0.86	Te, 9.9
Si (311)	9.5	22.96	2.4	0.64	Rh, 5.8
Si (311)	15	14.64	0.95	0.41	Kr, 2.8
Si (311)	65	4.182	0.03	0.12	Ca, 0.8
Si (111)	5	22.71	4.5	2.9	Rh, 5.8
Si (111)	7.7	14.77	1.9	1.9	Kr, 2.8
Si (111)	15	7.647	0.50	0.99	Co, 1.3
Si (111)	65	2.184	0.018	0.28	P, 0.5
Ge(111)	5	21.80	4.3	7.13	Ru, 5.3
Ge(111)	15	7.342	0.48	2.40	Fe, 1.3
Ge(111)	65	2.097	0.017	0.68	P. 0.5
InSb (111)	7.5	12.71	1.7	5.6	Se, 2.3
InSb (111)	12.5	7.665	0.60	3.1	Co, 1.3
InSb (111)	65	1.831	0.015	0.79	Si, 0.5

devices, so that the angular step becomes less obvious to the scientific user. The smallest angular steps are in the region of 1 μrad (0.2 arc s) and this can always be finer than the resolution limiting factors of the monochromator crystal and the absorption-edge features.

The second factor and third factors are convoluted together. The overall resolution relates to the square root of the sum of the squares of the beam divergence ($\Delta\theta$) and the Darwin width ($\Delta\theta_B$) of the crystal reflection (equation 3.9). The smaller source sizes and divergences in third-generation sources greatly reduce divergence due to the source itself. For SuperXAS at the SLS, the vertical source size (16 μm) and divergence (0.6 mrad) provide a much better basis for energy resolution. The incorporation of collimating mirrors also reduces the divergence so that the effective resolution becomes dominated by the Darwin width of the crystal, and any heat bumps on that crystal. An example of a bending magnet beamline is B18 at Diamond, which has water cooling. A temperature variation of 13° results from the beam on the first crystal. The slope errors are much smaller than Darwin widths for the Si(111) crystal at low energies, but are comparable to the Darwin widths in all the other cases. However, the resolution loss due to the heat bump is largely restored after the reflection from the second crystal since the heat load there is extremely small.

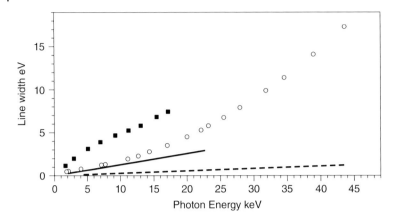

Figure 3.13 Darwin width of the a) Si(111) (solid line) and b) Si(311) (dashed line) crystals compared with the core-hole lifetimes (eV) of the *K* (○) and *L₃* (■) absorption edges as a function of photon energy (keV).

From Table 3.2 and Figure 3.13, we can compare the Darwin widths of crystals to the core-hole lifetimes, which limit the energy resolution of a standard XAFS scan. The resolution ($\Delta E/E$) derived from the Darwin width does not vary with Bragg angle but with crystal plane, estimated as 1.29×10^{-4} and 2.73×10^{-5} for the (111) and (311) planes of silicon, respectively. The Si(311) plane provides a resolution that is considerably higher than the lifetime broadening, but this will come at a cost of a lower photon flux on the sample associated with the narrower energy band pass. The Si(111) plane generally provides a line width comparable to that of the core-hole lifetime of *K* edges and considerably smaller than the linewidth of *L₃* edges. The intrinsic resolutions of the Ge(111) and InSb(111) monochromators are poorer than the line-widths, but they can provide good bases for EXAFS scans.

The choice of monochromator crystals does influence the potential resolution of a measurement and thus should be made with care. Si(111) is the one most commonly used since it provides a good working compromise of flux and resolution, but it is not always the optimum. Fortunately, beamlines often house two different crystal sets in the monochromator vessel and changing between them can be achieved without breaking the vacuum in that part of the beamline. The angular spread of the reflection ($\Delta\theta_B$) from a crystal reduces strongly with the harmonic number (*n*), according to equation 3.12. There is also angular offset between the reflectivity maxima of the fundamental and the harmonic, for example, between the Si(111) and (333) reflections. (There is no intensity in a Si(222) reflection due to crystal symmetry.)

$$\Delta\theta_{Bn} = \frac{\Delta\theta_{B0}}{(n+1)^2} \qquad (3.12)$$

This forms a basis of harmonic reflection via an *order-sorting monochromator*. In a double crystal monochromator, the two crystal faces can be detuned to a small extent (~20 µrad) off the top of the rocking curve, generally to ~ 50% of the maximum intensity. Provided this reduction is in the direction away from the harmonic then the discrimination for the fundamental will be increased to a very high value. This achieves harmonic rejection without the need for a mirror, but it is at the expense of 50% of the available monochromatic x-radiation.

There are also spurious and unwanted crystal reflections that can occur. That creates a contamination in the monochromatic purity over a small angular range. These multiple reflections are due to simultaneous fulfillments of the Bragg condition for different Miller indices, and can be mapped.[11] Unfortunately, this can cause a change in the recorded XAFS spectrum adding a sharp feature from the spurious reflection (a crystal glitch), which is not normalized away. In the soft x-ray region there can be substantial regions of energy that can be glitch free, by choosing a favorable azimuthal orientation of the crystal. But in the medium and hard x-ray regions glitches are hard to avoid. Minimizing them is generally achieved by having high homogeneity in the sample and in the beam in the vertical direction.[12]

A final aspect of monochromator design is focusing. In principle by bending a monochromator crystal (the second crystal in a double crystal monochromator) in the plane perpendicular to the beam direction can focus the collected monochromatic x-rays into a smaller horizontal spot (sagittal focusing). For a beamline using an undulator this is not necessary since the source is small anyway, but this can potentially increase the flux on a sample of mm size as compared to confining the beam via post-monochromator (exit) slits. The radius of curvature requires varies with the Bragg angle of the monochromator, so to maintain the size of focus during an EXAFS scan, the crystal bend must be changed. Generally monochromator designs have incorporated ribs on the non-illuminated side of the crystal to achieve this dynamical bending. This has a tendency to create differential bending between the ribbed and unribbed regions. The combination of this and the dynamical focusing makes creating a stable beam position and profile over a wide monochromator angle range a challenge. As a result, the preferred focusing on new bending magnet beamlines is by mirrors.

3.3.3   Near-Sample Focusing Elements

In order to achieve very small beams at the sample, near-sample focusing elements can be employed for a second stage of demagnification. The three main types are mirrors, lenses, and zone plates, each of which has different characteristics.

3.3.3.1   Kirkpatrick-Baez (KB) Mirrors
These are shown as the final pair of elements in the schematic of the beamline I18 at Diamond (Figure 3.4). Each mirror is bent into an elliptical shape with

a four-point bender, and focuses in orthogonal planes. They can be coated with stripes of different elements to allow for operation across the full beamline energy range. Their properties are non-chromatic. Hence the beam position does not vary across an EXAFS scan. Beam spot sizes of 1–3 µm are generally achieved, with recent reports extending this to sub-micron ranges, such as 350×700 nm for ID21 at the ESRF and 30 × 50 nm at BL29XUL of Spring-8, by reducing the size of the virtual source and the photon flux.[13] The working distance to the sample can be about 100 mm, giving options for *in situ* arrangements.

3.3.3.2 X-Ray Lenses

Parabolic refractive lenses have been made by exploiting the fact that the refractive index of a metal, for example, Be or Al, in the hard x-ray region is <1.[14] Hence drilling a circular hole, creating a concave shape for the metal, acts as a focusing lens. The focusing of one lens is very small as the difference in refractive index between the metal and air is only 10^{-3}–10^{-4}. However, a stack of N lenses will decrease the focal length by $1/N$. With parabolic-shaped lenses the spherical aberration was removed and focal lengths of about 1 cm could be achieved giving spot sizes in the region of 100 nm. The lens elements are then within 1 cm of the sample, reducing sampling options, but the focus is independent of the x-ray wavelength and such devices can be used for spectroscopy.

Lenses based upon reflective optics, called multilayer Laue lenses (MLL) have also been demonstrated as nanofocusing devices in the hard x-ray range (12–20 keV).[15] These devices are fabricated from lined structures of alternating WSi_2 and Si layers and crossing two of the devices gives focusing in two dimensions. The size of the alternating layers decreases away from the center of the x-ray beam from 25 to 5 nm. Spot sizes down to 25 nm have been achieved at the APS (Beamline 26-ID). The resolution and the focal position are dependent upon the x-ray wavelength, so XAFS applications are likely to be in x-ray emission or XANES.

3.3.3.3 Zone Plates

The progenitor of the MLL was the Fresnel zone plate (FZP). Like the MLL, there is a specific gradation of alternating layers of differentially absorbing material, though in this case it is in a series of concentric rings, again with the outermost layers having the closest spacings. In the condition when this spacing is larger than the wavelength, the maximum resolution is given by $1.22.\Delta r_N$, where Δr_N is the smallest (outermost) zone width. The distance from the plate to the focus (f) is also related to this feature, the radius of this zone (r_N) and the radiation wavelength by equation 3.13

$$f = \frac{2r_N \Delta r_N}{\lambda} \tag{3.13}$$

Advances in lithographic printing and in electrochemical deposition have resulted in the resolution of 15 nm lines.[16] The zone plate structure was created by electron beam lithography of a silsesquioxane resist on a silicon nitride membrane. On the structure, iridium was deposited by atomic layer deposition, although electrochemical deposition has also been used for this last step. The effect of the iridium is to double the frequency of the zone plate structure, thus reducing Δr_N. As in all cases of diffractive optics, the focal length of the focusing device varies with x-ray wavelength and it is difficult to make this imaging device compatible with EXAFS measurements, although other XAFS techniques can be applied.

3.4 Effect of Photon Energy on Experiment Design

A key factor in experimental design is the photon energy to be used. This is not just from the point of view of fashioning it on the beamline, but in the variation of the mass absorption coefficients of materials in and around samples. Four representative materials are shown in Figure 3.14.

The dotted line shows the x-ray transmission of 100 mm of air. From that it can be seen that for energies of less than 5000 eV (near the Ti K edge), a very significant proportion of the light is absorbed and thus any air path between source, sample and detectors should be minimized. For soft x-ray energies (<2 keV), transmission becomes minimal and hence air in and around the sample should be eliminated and replaced by vacuum or helium.

In the second curve 2 mm of water is taken as an exemplar for a solvent or material consisting of first row atoms. In this case, a 2-mm-path length is

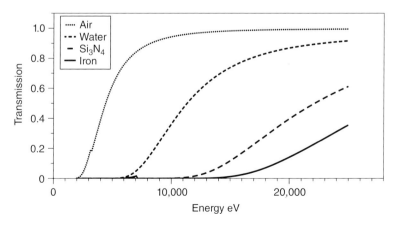

Figure 3.14 Transmission calculated through a sample of air (100 mm), water (2 mm), silicon nitride (1 mm), and 0.1 mm of iron (0.1 mm).

acceptable for energies over ~8 keV (Co K-edge) and at that point air absorption is considerably reduced and one can be less fastidious with vacuum pipes to the sample location. The transmission curve slopes strongly in the medium x-ray energy range and there is great advantage in optimizing the cell pathlength.

The longer dashed line is from 1 mm of Si_3N_4, representative of a second row element containing material; a powdered version of such a material packs to ~50% of the density of the pure materials, and so the curve relates roughly to a 2 mm depth of powder. Absorption is clearly very marked below about 14 keV (near the krypton K edge), so path lengths for studying elements with lower energy edges should be reduced below these values.

Finally a 0.1-mm-thick Fe sample is also shown as the solid line. Apart from a small absorption dip just below the Fe K edge such a path length would be problematical below 20 keV (Mo K-edge). This has consequences for the choice of window material. Use of 50 μm Fe windows implies a 100 μm source-to-sample-to-detector path length in transmission measurements and ~140 μm with fluorescence detection. This will cause a serious diminution of detected signal below ~26 keV. In addition, any measurements on a sample consisting largely of a first row transition element will only probe a depth of the order to 10 μm below about 15 keV, and much less near above the Fe edge itself.

With a degree of care with sample and cell design, a wide range of materials can be studies by XAFS and in the next chapter we concentrate on how to carry out these experiments.

3.5   Questions

1   **A** Calculate the E_c for the ESRF bending magnets under the two conditions that apply in that source (equation 3.1):

Electron energy = 6 GeV; Magnetic field = 0.4 and 0.8 T

B Bending magnet sources are generally used only up to about three times the E_c value. What is the reason for this?

C What is the highest practical energy that should be used on a Diamond bending magnet sources (Electron energy = 3 GeV; B = 1.4 T).

D What would this be if a wavelength shifter with a field of 3.5 T were installed?

2   Calculate the deflection factors, K, and the energy ranges of the harmonics up to $n = 15$ emanating if the following insertion devices were incorporated into Diamond (E = 3 GeV) (equations 3.4 and 3.5):

A A 30-mm period undulator with $B_{max} = 0.8$ T (at magnet gap 10 mm) (A)

B A 20-mm period undulator with $B_{max} = 0.8$ T (at gap 7.07 mm) (B)

C Calculate the outputs that would accrue if the gaps for devices A and B were interchanged.

D Which of these configurations would be appropriate for studying the following absorption edges: Si K (1.839 keV), Fe K (7.112 keV), Au L_3 11.919 keV) and Ag K (25.514 keV)?

3 **A** Calculate the monochromator angle change required to carry out an EXAFS scan at the molybdenum K edge (20.000 keV) using a Si(111) monochromator (start 300 eV below the edge and finish 700 eV above the edge).

B Calculate the resolution in eV that would be achieved with a 0.25 mdeg angle resolution on the monochromator.

C Calculate the resolution that would be achieved with a Si(311) monochromator, and indicate which one is preferred for XAFS measurements.

4 **A** Considering the reflectivity curves in Figure 3.9, which shows two different mirror coatings and incident angles, suggest a configuration to study the following elements by EXAFS.

   i) Cr K-edge (5.9 keV),
   ii) Fe K-edge (7.1 keV)
   iii) Au L_3 -edge (11.9 keV)

B What might be the value of platinum-coated mirror? Why may Rh have been chosen in preference to Pt for beamline I18 at Diamond (energy range 2 – 20 keV)?

References

1 Lightsources.org can be accessed on www.lightsources.org.
2 'X-ray Data Booklet', Ed. A. C. Thompson, Center for X-ray Optics and Advanced Light Source, LNBL/PUB-490 Rev. 3, 2009. (http://adb.lbl.gov/xdb-new.pdf). On-line calculations from http://henke.lbl.gov/optical_constants.
3 'Brilliant opportunities across the spectrum', J. Evans, Phys. Chem. Chem. Phys, 2006, **8**, 3045–3058.
4 'Generation of femtosecond pulses of synchrotron radiation', R. W. Schoenlein, S. Chattopadhyay, H. H. W. Chong, T. E. Glover, P. A. Heimann, C. V. Shank, A. A. Zholents, M. S. Zolotorev, Science, 2000, **287**, 2237–2240.
5 'Femtosecond XANES study of the light-induced spin crossover dynamics in an iron(II) complex', Ch. Bressler, C. Milne, V.-T. Pham, A. ElNahhas, R. M. van der Veen, W. Gawelda, S. Johnson, P. Beaud, D. Grolimund, M. Kaiser, C. N. Borca, G. Ingold, R. Abela, M. Chergui, Science, 2009, **323**, 489–492.

6 'Development of the hard x-ray source based on a tabletop electron storage ring', H. Yamada, Y. Kitazawa, Y. Kanai, I. Tohyama, T. Ozaki, Y. Sakai, K. Sak, A. I. Kleev, G. D. Bogomolov, Nucl. Instrum. Meth. Phys. Res. A, 2001, **467**, 122–125.

7 'Laser-EXAFS: Fast extended X-ray absorption Fine Structure Spectroscopy with a single pulse of laser-produced X-rays', P. J. Mallozzi, R. E. Schwerzel, H. M. Epstein. B. E. Campbell, Science, 1979, **206**, 353–355.

8 'Adaptive temporal and spatial shaping of coherent soft x-rays', C. Winterfeldt, T. Pfeifer, D. Walter, R. Kemmer, A. Paulus, R. Spitzenpfeil, G. Gerber, C. Spielmann, Proc. SPIE, 2006, **6187**, 61870 F.

9 Darwin widths: http://www.chess.cornell.edu/oldchess/operatns/xrclcdwn.htm

10 'Combining flat crystals, bent crystals and compound refractive lenses for high-energy x-ray optics', S. D. Shastri, J. Synchrotron Radiat., 2004, **11**, 150–156.

11 'Determination of glitches in soft X-ray monochromator crystals', G. van der Laan, B. T. Thole, Nucl. Instrum. Meth. Phys. Res. A, 1988, **263**, 515–521; http://ssrl.slac.stanford.edu/~xas/glitch/glitch.html.

12 'Minimizing "glitches" in XAFS data: A model for glitch formation', F. Bridges, X. Wang and J. B. Boyce, Nucl. Instrum. Meth. Phys. Res. A, 1991, **307**, 316–324.

13 'Development of scanning x-ray fluorescence microscope with spatial resolution of 30 nm using Kirkpatrick-Baez mirror optics', S. Matsuyama, H. Mimura, H. Yumoto, Y. Sano, K. Yamamura, M. Yabashi, Y. Nishino, K. Tamasaku, T. Ishikawa, K. Yamauchi, Rev. Sci. Instrum., 2006, **77**, 103102.

14 'Refractive x-ray lenses', B. Lengeler, C. G. Schroer, M. Kuhlmann, B. Benner, T. F. Günzler, O. Kurapova, F. Zontone, A. Snigirev, I. Snigireva, J. Phys. D: Appl. Phys., 2005, **38**, A218–A222.

15 'Two dimensional hard x-ray nanofocusing with crossed multilayer Laue lenses', H. Yan, V. Rose, D. Shu, E. Lima, H. C. Kang, R. Conley, C. Liu, N. Jahedi, A. T. Macrander, G. B. Stephenson, M. Holt, Y. S. Cu, M. Lu, J. Maser, Optics Express, 2011, **19**, 15069–15076.

16 'Ultra-high resolution zone-doubled diffractive x-ray optics for the multi-keV regime', J. Vila-Comamala, S. Gorelick, E. Färm, C. M. Kewish, A. Diaz, R. Barrett, V. A. Guzenko, M. Ritala, C. David, Optics Express, 2011, **19**, 175–184.

4

Experimental Methods

The two preceding chapters illustrate important characteristics about the framework in which experiments to be designed and executed. The vast majority of XAFS measurements are carried out on storage ring sources via applications for a limited amount of beam time. Generally there is no guarantee of a second chance should the experiments fail due to user error; the starting position of assessors will be: "Why didn't you do it right the first time?" Almost equally serious is not having sufficient beam time to complete the project. At best, this might mean a delay of 6 months in completion of a research paper or thesis. Accordingly, pre-planning of an experiment at the proposal stage is of high importance. This should have a scope covering:

1) Obviously, why is this scientifically more interesting than competing proposals?
2) Does the experiment match the characteristics of the source?
3) Do the facilities at the beamline accord with your needs?
4) Will any experimentation of your own gel with the space, control systems, and safety requirements of the facility?
5) Which absorption edges will you study? Can the beamline be easily configured for any changes that are in mind?
6) What is the best way of presenting the samples for your experiment? How long might each measurement take?
7) Do you need to record XAFS spectra of a set of reference materials to aid your analysis?
8) Including a modicum of contingency, what is the total time required to complete this experiment?
9) Is this viable given the staffing that you can provide at the experiment?

A careful analysis of this, with discussions with the beamline staff, will increase the probability of success with applications as well as experimental execution. Experiments are broadly in two classes: technique development and scientific applications. Defining the beam time for the former can be

X-Ray Absorption Spectroscopy for the Chemical and Materials Sciences, First Edition. John Evans.
© 2018 John Wiley & Sons Ltd. Published 2018 by John Wiley & Sons Ltd.

more difficult, and require more contingency. But the majority of applications are of the latter type when the majority of time is expended on creating the sample state and carrying out measurements on the resulting materials.

4.1 Sample Characteristics

4.1.1 X-Ray Absorption of Samples

The absorption properties of the samples are often the most important, and least considered, aspect of experimental design. Here, we amplify what was outlined in Sections 2.1 and 3.5. The key formulae are reproduced here, relating the x-ray transmission before (I_0) and after (I_t) the sample, the linear absorption coefficient (μ), the path length of the beam through the sample (l), mass absorption coefficient (μ_m), and density (ρ). For composite samples, absorbance $\mu(E)$ at an energy E is the weighted sum of the components.

$$I_t \Big/ I_0 = e^{-\mu l} \tag{4.1}$$

$$\mu(E) = log_e \left(I_t \Big/ I_0 \right) = \Sigma_i w_i l_i \left(\mu_m / \rho \right)_i \tag{4.2}$$

The *attenuation length* (λ) of a sample is the path length at which the x-ray transmission drops to $1/e$ of the incoming value, where e is the mathematical constant and is of value ca. 2.71828, and occurs at:

$$\lambda = 1 \Big/ \mu(E) \tag{4.3}$$

The attenuation length is a useful guideline for considering sample absorption as it represents a transmission of about 37%. In practice it is advisable where possible to keep the background absorption below a value for $\mu(E)$ of 1 (i.e., at most 10% of the light may reach the detector after the sample), which will occur at ~2λ.

The attenuation of air is a fundamental factor in atmospheric design. From Figure 4.1, it can be seen that this only passes above 2.5 cm (an inch being a reasonable rule of thumb!) at ~2.4 keV, near the energy sulfur K edge. It is clear that air absorption should be avoided up to these energies. By the calcium K edge (4 keV) the attenuation length increases to ~1 dm and so small air gaps between beam pipe, sample and detectors can be accommodated. Only at ~8 keV (Ni K edge) does air absorption reach a value that might be neglected.

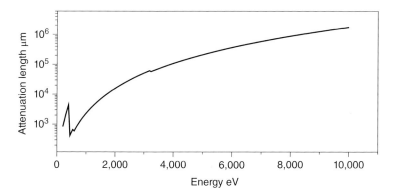

Figure 4.1 Attenuation length of dry air as a function of energy (250–10,000 eV).

4.1.2 Classes of Experimental Layouts

The most challenging experiments to set up are in the *soft* or *low-energy x-ray* region. As can be seen from Figure 4.1, photon energies below 2.5 keV (S *K* edge) are increasingly likely to be compromised by any significant air path between the components of the beamline and also with the sample cell. Generally, measurements are carried out with the sample *in vacuo*, ideally on a windowless beamline, that is, one with no window between the electron beam and the sample. Differential pumping is used to minimize the risk to the storage ring vacuum, but contamination in the vacuum vessel is to be avoided. Thus most soft x-ray XAFS measurements are carried out on involatile materials mounted on a manipulator allowing several samples to be studied before venting the sample chamber. Volatile samples or *in situ* studies can be carried out in custom made cells with appropriate windows, often polyimides with good physical and chemical stability. Figure 4.2 presents the x-ray transmission of a typical, thin film (25 μm of a polyimide). At the P *K* edge (2146 eV) such a film will transmit about 30% and thus from this energy point the window does not present a serious difficulty. A more severe problem is encountered at the Mg *K* edge (1303 eV), where the window transmission will be reduced to 0.009. *In situ* studies at lower energies will require either extremely thin carbon windows or carefully designed windowless cells with a thin layer reagent above the surface of interest.

In the *medium-energy x-ray* region (~4 to ~10 keV) the constraints of air absorption lessen very significantly and thus a laboratory-style spectroscopic experiment can be laid out on an experimental station. Window materials composed of light elements (e.g., plastics, boron nitride, carbon) can be used throughout. Thin windows of second row elements (Al, quartz) become viable, and, as exemplified in Chapter 6, the nature of the material under study itself sets the optimum for sample presentation. In the *hard* or *high-energy x-ray* region (>10 keV), these constraints lessen further and there is great flexibility for experimental design.

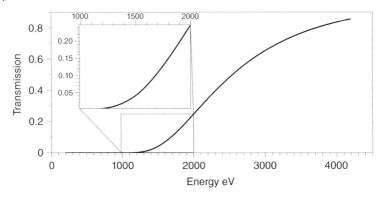

Figure 4.2 Transmission of a polyimide film (25 μm) between 200 and 4200 eV.

4.2 Scanning Modes

4.2.1 Scanning XAFS

The normal mode of recording a XAFS spectrum involves moving the Bragg angle of the monochromator from high to lower angle in a stepwise fashion, using a system outlined in Section 3.3.2 and illustrated in Figure 3.4. There may be a dwell time at that angle to allow resetting of optics and positioning prior to the acquisition time at that energy. The scan range and interval may be set either in angle (millidegrees) or, by means of an encoder on the beamline, directly into energy (eV). From equation 3.10 and a Si(111) monochromator (d = 3.1356 Å) the relationship between angle and energy can be written as in equation 4.4.

$$E = \frac{12.3984}{2d\sin(\theta)} = \frac{1.977}{\sin(\theta)} \, keV \qquad (4.4)$$

A simple stepwise acquisition with a set interval and acquisition time is satisfactory for short test or edge scans, but it is not the most efficient procedure for recording EXAFS or carrying out kinetic studies. A full EXAFS spectrum can be split into three regions (Figure 4.3). The pre-edge region is acquired to provide a reliable baseline for background subtraction and thus can be acquired with a relatively large interval (5–10 eV) over enough energy range (200–300 eV) to provide a good extrapolation across the EXAFS spectrum and a relatively short acquisition time per point. The edge region requires a much finer step interval (~0.25 eV), generally to ensure that natural line-width of the XANES and pre-edge features are not distorted by the digitization interval. These features are relatively sharp and intense and so the acquisition time per point need not be increased very substantially, unless there are weak pre-edge features. The third region is generally analyzed as EXAFS in *k* space, from

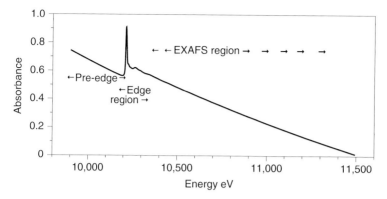

Figure 4.3 Regions of a XAFS spectrum, shown for the W L_3 edge of $(NBu_4)_2[W_6O_{19}]$ in CH_3CN solution (*Source:* Diamond, B18, data from Richard Ilsley).

about $k = 3$ Å^{-1} ($\sim$34 eV above E_0) until the end of the recorded range ($\sim$12 to $\sim$20 Å^{-1}). The square relationship between $(E–E_0)$ and k (equation 4.5), means that keeping a constant interval in k ($\sim$0.025 Å^{-1}) can be achieved with an increasing energy interval.

$$E - E_o = \frac{k^2}{0.262449} \ eV \tag{4.5}$$

But offsetting this is the dampening of $\chi(k)$ with increasing k. EXAFS analysis is generally on data with a k weighting $k^n.\chi(k)$, where $n = 1$, 2, or 3. Given ideal statistics, the signal to noise ratio (S/N) improves as the square root of the acquisition time (t) (equation 4.6), retrieving the S/N over a the EXAFS data range with k^1 weighting would of itself require the acquisition time to have a weighting of k^2. Hence the acquisition time per data point should also increase across the spectrum. Ideally this should be smoothly using a k^n weighting ($n \sim 3$), but if that algorithm is not available it can be approximated by dividing the EXAFS range of the spectrum into regions with increasing energy interval and acquisition time per point.

$$S/N \propto \sqrt{t} \tag{4.6}$$

This mode is appropriate for long scans on dilute materials when the acquisition time is considerably longer than the dwell time and thus the duty cycle is very efficient (i.e., a low dead time between points). An alternative method for spectra that can be usefully recorded more quickly is the Quick EXAFS mode (QEXAFS). In that the monochromator is driven at a constant velocity and data is acquired on the fly. This can be effected in both forward and backward directions. There is then no dwell time and so the acquisition duty cycle is extremely efficient. Care must be taken to check the calibration of both forward and reverse scans, which may show a hysteresis effect. However, this mode has been successfully enacted on many facilities providing a scan time of the order of seconds. In its most rapid mode, where the oscillation of the monochromator crystal is achieved using an eccentric cam system, XANES spectra have been successfully acquired in 25 ms, and EXAFS in under 1 second.[1]

4.2.2 Energy Dispersive XAFS

In this arrangement, one crystal is used to select x-ray energies in a horizontal, deflecting plane, but the bandwidth is increased by bending the crystal to provide a spread of Bragg angles. The effect of this is to focus the x-ray beam on the sample and then disperse the x-ray bandwidth onto a multi-element linear detector (Figure 4.4). Generally a Bragg monochromator is used and the crystal bent to a curvature that will provide XANES and EXAFS at one absorption edge, although neighboring edges can also be measured simultaneously.

The main advantage of the dispersive configuration is that of multiplexing. The acquisition time will be similar to that of a single energy point in a scanning spectrometer, which can be less than 1 ms. However, the technique is very demanding on x-ray beam stability, especially since the background spectrum, I_0, is not acquired synchronously, as is the case in a scanning spectrometer. Its applications tend to be situations where small beam-size and rapid scan times are essential, such as in kinetic studies, samples at high-pressure experiments in diamond anvil cells, and in sample mapping.[2]

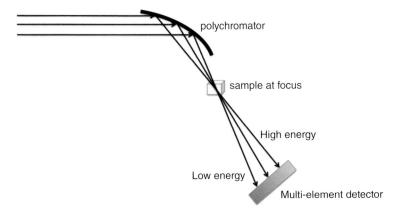

Figure 4.4 Schematic representation of energy dispersive XAFS using a Bragg monochromator.

4.3 Detection Methods

4.3.1 Transmission

This is the standard reference method of detecting x-ray absorption directly. Detectors placed before and after the sample provide a relative absorbance. They need to have the properties of high resilience to x-radiation, high sensitivity, and excellent linearity. Ion chambers are commonly used. The absorption of the device can be tuned by varying the pressures of the inert gases in the chamber. Typically the first ion chamber (I_0) should have sufficient absorption (~10 to ~40%) to provide a strong signal but allow the majority of the light to pass through to the sample. The second chamber (I_t) has a more absorbing mix (~80 to ~95%) leaving the residue to pass through to a reference sample, typically an appropriate metal foil, with a third ion chamber (I_r) downstream of it. In this way the second and third ion chambers provide the signals for the spectrum of the reference material (Figure 4.5a). The ion chamber (Figure 4.5b) has parallel plates with a high potential difference and acts as an integrating device with no energy resolution. The voltage (~300 V) is set so that the cations from the ionized gas and the electrons travel to the collector plate rather than recombine and the current generated is linear with the x-ray flux. At higher voltages the electrons generated can also ionize the gas and the characteristics of the chamber change from a highly linear converter of photon flux to current, through a non-linear regime, to a proportional counter where the electron multiplication effect creates avalanches. In the ionization chamber region, the ion current, I_{ion} is given by equation 4.7, showing the proportionality

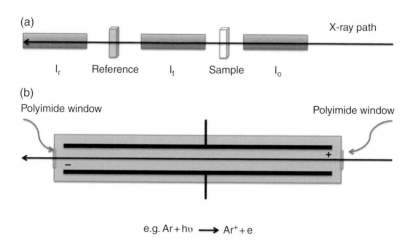

Figure 4.5 a) A configuration for measuring XAFS spectra in transmission and b) the components of an ion chamber.

to the photon flux absorbed in the ion chamber (N) and also to the photon energy (E_{ph}). There is a degree of amplification from ratio of the photon energy with that required to create the electron-ion pair.

$$I_{ion} = N.e\left(\frac{E_{ph}}{E_{gas}}\right), \text{ where } e = \text{electron charge} \tag{4.7}$$

For argon $E_{gas} = 26$ eV, and as a result each photon of energy $E_{ph} = 10$ keV will create 385 electron-ion pairs. Figure 4.6 presents the absorption efficiency of four gases often used in ion chambers, calculated for a 0.5 bar pressure and a 10 cm path length. It is evident that argon and krypton can be used to provide absorption efficiencies over these energy ranges by varying the partial pressures. Helium may be employed as essentially a relatively transparent filler gas; nitrogen is a viable ionizable gas in the soft x-ray region. However, as well as limitations in the applied voltage, there are also problems created by the high photon fluxes created by insertion devices on third-generation sources. The maximum count rate is in the order of $10^{11}\text{phs}^{-1}\text{cm}^{-3}$. Reducing the absorption by having first row elements in the ion chamber is one approach to offsetting this problem; not only does it reduce the absorption coefficient of the gases, it also tends to increase E_{gas} (36 and 41 eV for nitrogen and helium, respectively).

The alternatives to ion chambers are principally thin semiconductor devices operating in a current mode. X-radiation creates a hole-electron pair, which requires much less energy than the current-forming process in an ion chamber (e.g., 3.6 eV in silicon), thus the current yield per photon is higher (2780 at $E_{ph} = 10$ keV). Doping of an *intrinsic (i)* semiconductor like silicon, with a Group 15 element, P, As, or Sb, results in an extra electron in the conduction band per dopant atom and is thus termed *n-type (negative)*; the corresponding

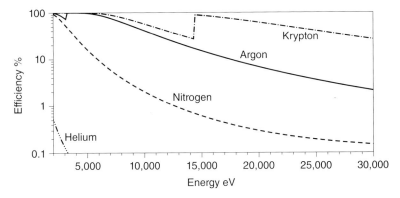

Figure 4.6 The absorption efficiency of a 10 cm path length of He, N_2, Ar, and Kr at 380 torr pressure (2–30 keV).

effect with a Group 13 element (B, Al, or Ga) creates holes in the valence band and is termed *p-type (positive)*. A *p-n* junction, as in most photovoltaic cells, results in a space-charge region whereby there is a flow of electrons from the *n* material to the *p* side thus reducing the number of conductors in a depletion layer formed at the junction. This depletion layer can, in effect, be expanded by adding a sandwiched layer of the *intrinsic* semiconductor in what is known as a *pin diode*. The effect of the reverse bias voltage on a diode is shown in Figure 4.7. Starting from a *p-n* junction, the diffusion of electrons to create the depletion layer causes a bend in the energy levels of the valence and conduction and bands. The potential gradient is increased with a reverse potential and this can be set to minimize the capacitance of the diode, indicating the depletion region is at a maximum and the single/background can be optimized.

However, the absorption properties of the photodiode are set by the material and its thickness. As can be seen from Figure 4.8, thin diodes (<10 µm) are needed for measurement of I_0 at low energies. A crystalline diode in that position will also generate diffraction and this will reduce the intensity of the through beam, creating sharp glitches in the apparent I_0 value. Generally, this will exacerbate normalization of the spectrum, especially for dilute and/or inhomogeneous samples. Saturation at high count-rates and flux densities can also be problematical. An alternative is to place a photodiode off the light path so as to intercept a small fraction of the straight-through beam scattered off a foil or even from air, but this will reduce sensitivity.

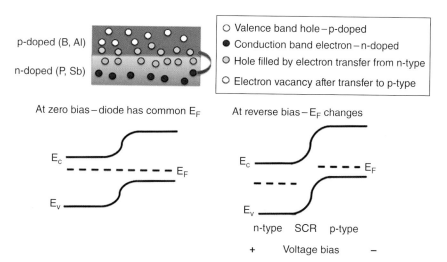

Figure 4.7 Formation of a depletion layer in a *pn* junction of silicon-based semiconductors and the effect on the energies of the top of the valence band (E_v), the Fermi level, and the bottom of the conduction band (E_c) due to electron diffusion and a reverse voltage.

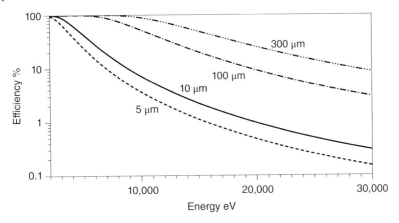

Figure 4.8 Absorption efficiency of silicon at different thicknesses (2–30 keV).

The sample itself has much to do with the quality of the spectrum recorded. In general, the absorption should be sufficiently large to give a significant change in the transmitted x-ray intensity (I_t). However, if the absorption is too high then there are too few photons reaching the detector and a high noise level will ensue. If a front ion chamber (I_0) is set to absorb 20% of the incident light (close to the optimum), and the sample has an absorbance ($\mu.l$) of 1, then from equation 4.1, we can calculate the flux reaching the second ion chamber (I_t). If the second ion chamber is set to absorb 90% of the incident beam, then its sensitivity will be 4.5-fold higher. As a result we can plot the relative signal intensity on I_t/I_0 (Figure 4.9a). Clearly a μ value >5, which transmits <0.6% of the x-rays incident on the sample, will compromise the signal intensity on the second ion chamber very considerably. On the other hand, a very small change in absorbance with $\Delta\mu$ <0.01 will cause a change in I_t of <1% and again be non-optimum. In transmission mode the signal/noise ratio generally improves in way that is close to the information theory relationship, that is, to the square root of number of photon counts $\sqrt{N_{ph}}$. Accordingly, the relative S/N can be estimated (Figure 4.9b). This shows an optimum range between 2 and 3 units, corresponding to a transmission through the sample of between 4 and 5%.

The sample thickness to achieve this varies with the chemical composition of the material, its density and the absorption edge energy (Figure 4.10). Taking twice the attenuation length (2λ) as a rule of thumb, then a uniform film should be created, which will vary from about 8 μm for the sulfate to 140 μm for molybdate. This is not trivial starting from typical powdered samples, and errors in sample presentation can result in distorted relative intensities. A danger is that an inhomogeneous sample is presented to the x-ray beam, with variable sample thicknesses (Figure 4.11). As shown in Figure 4.12, the result of this can be an increase in the transmission to the second ion chamber. The additional light, however, is not passing through the sample and contains

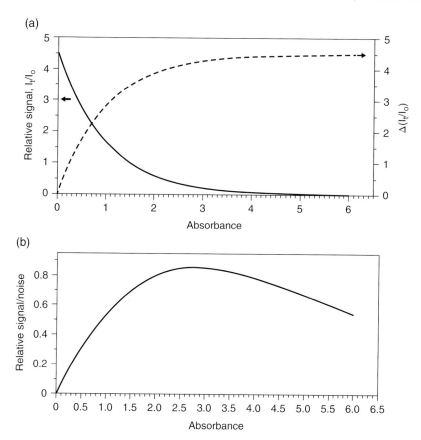

Figure 4.9 a) Relative signal intensity on I_t and I_0 ion chambers as a function of sample absorbance, μ, if 90% and 20% absorbing, respectively. b) Relative signal/noise (S/N) versus sample absorbance.

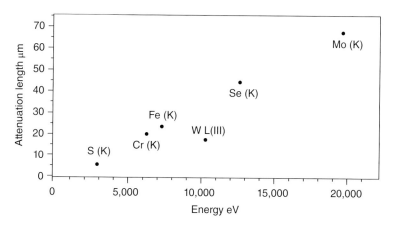

Figure 4.10 Attenuation length (μm) at 50 eV above the absorption edges of compounds of formula Na_2EO_4 at normal density (E = S, Cr, Fe, W, Se, and Mo).

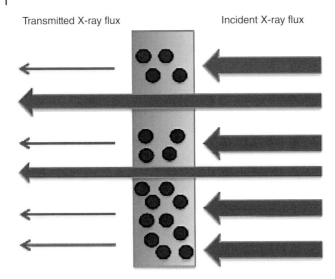

Transmitted X-ray flux   Incident X-ray flux

Figure 4.11 Effect of pinholes on the x-ray transmission through a solid sample.

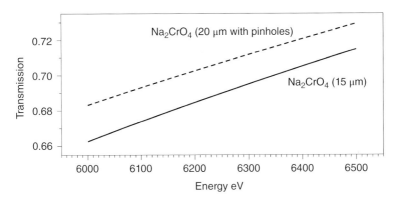

Figure 4.12 Comparison of the transmission of a sample of Na_2CrO_4 above the Cr K edge (ignoring EXAFS features) for a sample with a uniform thickness (15 μm) packed at a density of 1.4 gcm^{-3} with one of 20 μm thickness with 25% of the sample area containing pin holes.

no information about it. Hence the XAFS features will have an inaccurate intensity profile. Since these types of samples are often used for model structures, the risk is that this error is transferred to other solid samples or to solutions. Additional risks from inhomogeneous samples are that sample settling, and beam movement can vary the quantity of sample that is with the x-ray beam during the recording of the spectrum.

Solutions provide an excellent means of presenting a homogeneous sample and thus circumventing this problem. However, the starting points of many analyses are the reported crystal structures of model compounds. The normal means of approaching solution-type homogeneity with a true solid sample is to use a diluent and create a finely divided physical mixture, by, for example, mortar and pestle grinding or ball-milling. The diluent would generally be an inert, low atomic number material such as boron nitride, cellulose, or synthetic polymer powders. They may then be either laid into a sample holder of known thickness, or, more often, pressed into a pellet. If this is carried out in a glove box with a dried diluent, and the sample sealed between windows, then the technique can be applied to air-sensitive materials.

We can consider Na_2CrO_4 as an example for optimizing sample preparation. From equation 2.2 (*v.s.*), the quantity of Na_2CrO_4 (MW 166) required to give an absorption of 2 at a point sufficiently above the edge to be near a normalization level (50 eV, i.e., 6040 eV photon energy) can be determined from equation 4.8.

$$\log_e\left(I_0/I_t\right)=2=\Sigma\left\{w.l_{Cr}.\left(\mu_m/\rho\right)\right\}_{Cr+}\left\{w.l.\left(\mu_m/\rho\right)\right\}_{Na+}\left\{w.l.\left(\mu_m/\rho\right)\right\}_{Na}$$
$$=\left[\left\{0.337.\left(508.9\right)\right\}+\left\{0.277.\left(65.66\right)\right\}+\left\{0.386.\left(16.89\right)\right\}\right]mass/area$$

(4.8)

Hence mass/area needed = $2/196.4$ g/cm^2 = 10.2 mg/cm^2

The edge jump that would result from this sample can be calculated from the mass absorption coefficients below (63 cm^2/g) and after (519 cm^2/g) the edge. The mass weighted increase in the absorption coefficient is then 153.9 cm^2/g Na_2CrO_4, which, on this density of sample, would give an edge jump of ~1.6. The diluent and the sample thickness can be varied to ensure that this density of material is presented homogeneously to the beam. This combination of total absorption (2) and edge jump (1.6) should provide the basis of recording a high-quality spectrum; this should also be the case if the sample density were 50% higher than this. Generally, transmission should be the method of first choice, as it will in most cases give the best S/N and reliable spectrum. Indeed, recording in transmission has become viable to ever lower sample concentrations as sources become brighter and optics and detectors uprated; the threshold varies greatly with photon energy, with higher energies more favorable, and also with sample type (favoring lighter non-absorbing elements in the sample matrix). However, there are samples where transmission is not viable, and indirect measurement of absorption must be employed.

The effective sensitivity of transmission XAFS depends upon both upon the sampling depth and the intrinsic absorption change at an edge. Plots of the attenuation lengths for edges of elements from P to Po, and energies below 30 keV, are given in Figure 4.13 for solutions of 50 mM in aqueous solutions. Clearly for the higher edge energies, the greatly increased potential path length can improve sensitivity, providing there is sufficient sample for the attendant volume increase.

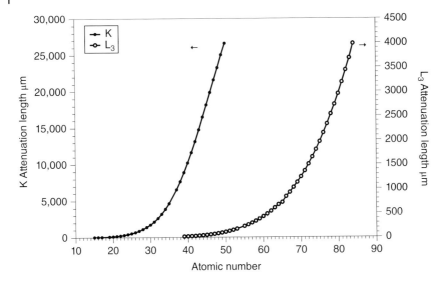

Figure 4.13 Attenuation length of aqueous solutions (50 mM) of elements (P –Po) at the onset of their K and L_3 absorption edges of energy less than 30 keV.

The resulting changes in sample absorption (edge jumps) at the sample attenuation length (Figure 4.14a) show the expected increases with atomic number. But it is also apparent that the relative slopes differ from those of the plots of attenuation length (Figure 4.13). Indeed, when this is normalized to the same path length (Figure 4.14b), it can be seen that the intrinsic sensitivity decreases with atomic number. It is also evident that the sensitivity of the L_3 edges is higher than that of the deeper-lying K edges; this is important for thin samples. Ultimately though, as sensitivity is pressed, the absorbance change of EXAFS oscillations in particular is too small compared to the discrimination of the detection system. In that situation, indirect methods may provide increased signal/background by virtue of having very low background signals.

4.3.2 Electron Yield

One of the main indirect methods of measuring x-ray absorption is to measure the current generated at the sample. The primary electron source emanates from the Auger electron emitted from that mode of relaxation of the core hole. But many more electrons come from the decay via secondary processes of photon and electron emission from relaxed states result in the much of the photon energy being converted into a cascade of ionization and inelastic events. A drain current the total electron yield (TEY) can then be taken from a sample holder and amplified to afford a XAFS spectrum with a linear relationship between absorption and drain current $\{i(E)\}$ (equation 4.9).

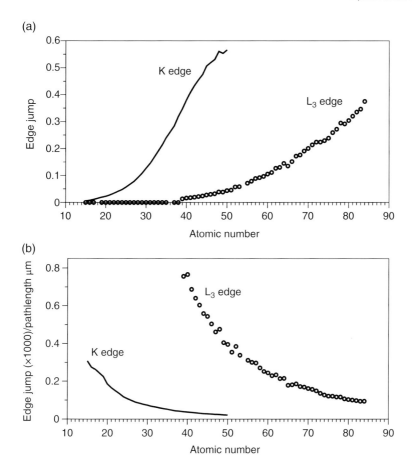

(a)

(b)

Figure 4.14 The change is x-ray absorbance of aqueous solutions (50 mM) of elements (P–Po) at their *K* and L_3 absorption edges of energy less than 30 keV. a) for a sample thickness of the attenuation length and b) per μm pathlength.

$$\mu(E) = Const. \frac{i(E)}{I_0} \tag{4.9}$$

This method can provide high sensitivity, but the XAFS component of the signal may be superimposed on a high background. As well as being a source of background noise, the background itself may fluctuate due to charging effects at the sample and its interface with a conducting mount. For insulating solid samples, this may be alleviated by intimate mixing with a conducting, involatile solid such as graphite.

Much of the use of electron yield detection is for the shallow absorption edges of light elements, for which the fluorescence yield is very low, and the

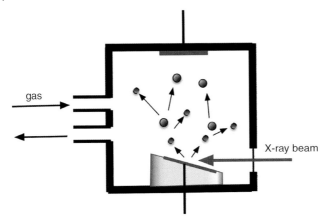

Figure 4.15 Conversion electron/ion yield detection.

emitted radiation has a low attenuation length exacerbating sensitivity issues. The signal/background ratio of a particular edge may be enhanced by using an electron energy analyzer to isolate the signal from electrons directly associated with the element of choice, for example by monitoring the Auger electrons (the Auger electron yield: AEY). The escape depth of the Auger electrons will vary with their energy, and this can be utilized to select a depth of interest in the region of a few nm. Rather than having a sample in a vacuum there may be an ambient gas, which is itself ionized by the electrons emitted from the sample. That creates a cascade of electrons and ions generating a conversion electron or ion yield (CEY or CIY) (Figure 4.15, adapted from [3]). This amplified yield may be detected as a current also providing a XAS spectrum according to equation 4.9. There is enhancement of the selectivity of the surface region as compared to a fluorescence detection method. The gas can also be a reactive one and also be at elevated pressures[4] and with a gas microstrip detector sensitivity is extremely high.[5]

4.3.3 Fluorescence

Fluorescence is the most common detection mode for dilute samples. In some sampling designs transmission is not possible, but fluorescence is generally employed when the edge jump is very small (and possibly also on a strongly sloping background). The XAFS features become an extremely small proportion of the total detector signals in the transmission spectrum and thus subject to significant noise. With a low background signal at x-ray energies below the absorption edge, fluorescence can then provide a better signal/background ratio. However, to achieve this benefit, some energy selection is necessary. This problem is illustrated in Figure 4.16. The most intense peak is from scattering

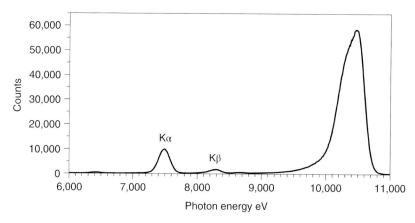

Figure 4.16 Counts measured from a nickel-containing sample with a multi-element germanium detector at 90° to the x-ray path (*Source:* Data from I20, Diamond).

of the incoming x-ray beam. At the highest energy is the elastically scattered peak, with inelastic effects (Compton scattering) due to momentum transfer from incoming x-ray to the electrons in the sample. The two smaller peaks are the $K\alpha$ (7.49 keV) and $K\beta$ (8.25 keV), the intensity of which can be proportional to the x-ray absorption, under the right sampling conditions.

The partitioning between Auger electron and x-ray emission moves to favor fluorescence with increasing atomic number, and thus absorption edge energy (Figure 4.17: data from [6,7]). The crossover of 50% probability for K emission is from Zn to Ga, but for emission from the L_3 edge this point is only reached for transuranic elements. The actual fluorescence yield available for detection is a function of the edge jump, fluorescence yield, and the absorption at both the incoming radiation and the fluorescence channels; the latter will have a shorter attenuation length than the incoming radiation, thus reducing the sampled volume. Each type of sample should be assessed on its own parameters, but the following examples can show some trends.

Samples containing germanium (K edge 11.11 keV) and iridium (L_3 edge 11.21 keV) will have similar attenuation lengths for the most likely edge of study. Both are convenient edges of study by XAFS, with the energy gap between the L_3 and L_2 edges of iridium being over 1600 eV. At their attenuation lengths lengths (2.63 mm for Ge and 2.33 mm for Ir), the edge jump of our reference 50 mM aqueous solutions would be 0.17 and 0.27 for Ge and Ir, respectively, in spite of the 13% larger sampling volume for the Ge sample. This higher edge jump for Ir (Ir/Ge = 1.6) is more than compensated by the lower fluorescence yield (Ir/Ge = 0.55), indicating an overall signal ratio of Ir/Ge = 0.9, roughly the ratio of the sample thickness. The reduction in relative sensitivity, however, is amplified by the fact that the α and β fluorescence channels are of

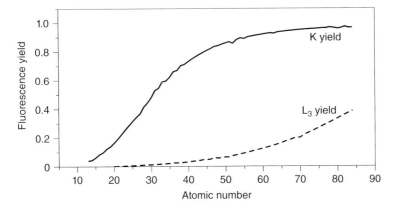

Figure 4.17 The fluorescence yield following absorption at the *K* and L_3 edges as a function of atomic number.

lower energy for iridium *L* (*Lα* 9175, 9099 eV; *Lβ* 10708 eV) than those for germanium *K* (*Kα* 9886, 9855 eV; *Kβ* 10982 eV). Accordingly the attenuation lengths of the emissions for iridium (for the 50 mM aqueous solution: at *Lα* 1.28 and 1.25 mm; at *Lβ* 2.03 mm) will be shorter than those for germanium (at *Kα* 1.86, 1.84 mm; at *Kβ* 2.56 mm); this will reduce the relative number of photons escaping the sampling environment for iridium rather than for germanium. This trend will pertain for comparisons between *3d* and *5d* transition series.

The second comparison is between the *K* and *L* shells of a *4d* transition element, taking as an example zirconium, the most abundant element of this transition series in the solar system. The energies of the *K* and L_3 edges are 17998 and 2223 eV, respectively. Hence the attenuation lengths of our sample 50 mM aqueous solution are very different: 21.4 µm and 10.192 mm before the L_3 and *K* edges, respectively. In this case then, the comparisons are very clear. At the sampling depth of the attenuation length, then the ratio of the edge jumps strongly and favors the higher energy edge ($K/L_3 = 24$), and this is amplified when considering the relative fluorescence yield to give a relative sensitivity of $K/L_3 \sim 600{:}1$. For thin samples the higher absorbance change per µm at the L_3 edge is compensated by the higher fluorescence yield of the K emissions. Given that the energies of the *Lα* (2042, 2040 eV) and *Lβ* (2124 eV) will make sample, window, and air absorption very high, and the energy range to the Zr L_2 edge is 84 eV, it will be a rare experiment for which the L_3 is favored over the *K* edge. The only advantages of the *L* edges will be the lower core hole line width (1.57 eV for L_3 compared to 3.84 eV for the *K* edge), and the differing transitions from the *2p* core electron that will both give valuable information in the XANES region.

4.3.3.1 Total Fluorescence Yield

The standard geometry for fluorescence detection (Figure 4.18) is designed to optimize the fluorescence signal over the scatter background, by placing the detector away from the x-ray path. The sample is angled close to 45° to the incoming beam in the horizontal and the effective sample thickness is increased by a factor of √2. The horizontal geometry is chosen since the synchrotron radiation is linearly polarized in that plane and scattering is minimized perpendicular to the incoming beam in the plane of polarization. The x-ray fluorescence is spread over the full solid angle (4π sr), and thus the geometry in Figure 4.18 normally provides the optimum I_f/scatter ratio. The global spread of the fluorescence signal means that the collection efficiency is generally limited by the solid angle subtended by the detector from the sampling point.

The energy difference between the absorption edge and the most intense of the consequential emission lines is about 10% and 20% for K and L_3 edges, respectively.

An ion chamber detector, as used in transmission measurements cannot provide the resolution required to distinguish scatter from fluorescence. Inserting a filter between the sample and the detector, can provide this to a significant extent. For many elements, the K edge of the preceding $(Z\text{-}1)$ element in the periodic table lies between the energies of the absorption edge and the $K\alpha$ emissions (Table 4.1). In that case the mass absorption of the filter will be low at the $K\alpha$ energy allowing the fluorescence to pass with little attenuation, but it will absorb a high proportion of the exciting radiation, thus increasing the I_f/background ratio (equation 4.10). Thus the signal/noise ratio can be improved if the background signal I_b is decreased or the acquisition time, t, is increased.

$$S/N = \left(\frac{I_f \cdot t}{1 + I_b / I_f} \right)^{1/2} \tag{4.10}$$

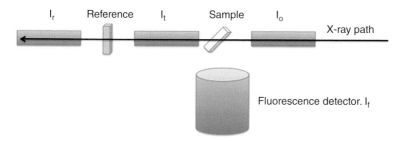

Figure 4.18 Configuration for measuring XAFS in fluorescence in addition to transmission.

Table 4.1 Energies (keV) of edges and emissions of the *K* absorption of the elements of the first long period and of potential Z-1 filters for fluorescence measurements. Gray shading indicates degree of problems monitoring with the emission line using the suggested filter.

Element	*K*	*Kα₁*	*Kβ₁*	Filter	*K*	*Kα₁*	*Kβ₁*
Calcium	4.04	3.69	4.01	Potassium	3.61	3.31	3.59
Scandium	4.49	4.09	4.46	Calcium	4.04	3.69	4.01
Titanium	4.97	4.51	4.93	Scandium	4.49	4.09	4.46
Vanadium	5.47	4.95	5.43	Titanium	4.97	4.51	4.93
Chromium	5.99	5.41	5.95	Vanadium	5.47	4.95	5.43
Manganese	6.53	5.90	6.49	Chromium	5.99	5.41	5.95
Iron	7.11	6.40	7.06	Manganese	6.53	5.90	6.49
Cobalt	7.71	6.93	7.65	Iron	7.11	6.40	7.06
Nickel	8.33	7.48	8.26	Cobalt	7.71	6.93	7.65
Copper	8.98	8.05	8.91	Nickel	8.33	7.48	8.26
Zinc	9.66	8.64	9.57	Copper	8.98	8.05	8.91
Gallium	10.37	9.25	10.26	Zinc	9.66	8.64	9.57
Germanium	11.10	9.89	10.98	Gallium	10.37	9.25	10.26
Arsenic	11.87	10.54	11.73	Germanium	11.10	9.89	10.98
Selenium	12.66	11.22	12.50	Arsenic	11.87	10.54	11.73
Bromine	13.47	11.92	13.29	Selenium	12.66	11.22	12.50
Krypton	14.33	12.65	14.11	Bromine	13.47	11.92	13.29

A favorable example would be the use of a copper foil as a filter for the Zn *Kα* fluorescence (Figure 4.19). The sample is modeled as a thin Zn sample with a transmission change at the edge from 0.97 to 0.83. The thicker copper foil (10 μm) has a transmission change from 0.72 to 0.078 at its edge. At the zinc *Kα* energy, the Cu foil would provide a transmission of 0.69, thus removing about 30% from passing to the detector. For the *Kβ* line, the filter will attenuate the signal by 0.89. The scatter over the whole of a likely pre-edge range and a full EXAFS run will be attenuated significantly (90–82%). The I_f/scatter ratio will have improved by this filter as the scatter has been attenuated much more, but the problem has not been eliminated. The scatter can be attenuated more with a thicker foil (by 95% for a 25 μm thickness), but at the expense of reducing the transmission of the *Kα* emission (to 40%).

For other elements, there is more difficulty in finding an effective filter (Table 4.1); for elements early in the first long period, there is no capacity for creating an unperturbed pre-edge or edge structure. The *Z-1* formula does not

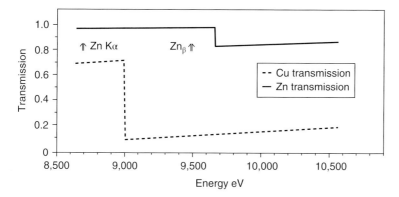

Figure 4.19 X-ray transmission of a 1 μm zinc "sample" with the effect of a 10 μm copper foil as a filter.

hold across for the L_3 edges of the third long period (Table 4.2). However, gaps can be filled by exploiting the K edges of lighter elements and, for a few metals, the $L\beta$ as well as the $L\alpha$ channel can be enhanced significantly over the background scatter.

These selectivity gains, however, are partially ameliorated by emissions from the filter (Tables 4.1 and 4.2). To reduce that effect, a set of collimating slits (Soller slits) can be mounted between the sample and the detector[8] (Figure 4.20). The fluorescence from the filter foil will be spread over the entire solid angle and so the majority of this unwanted background can be trapped by the collimating slits while allowing the majority of the fluorescence on the path from the sample to the detector to continue. In the example in reference,[2] Fe in a photosynthetic bacterial sample, the effective gain of the filter and the filter-slit assembly were estimated to be 2.6 and 6.4, respectively. In practice the gain was higher since the reduction in the background allowed the sample-detector distance to be reduced without saturation of the detector, and thus a larger solid angle of fluorescence could be collected.

As for transmission measurements (Section 4.3.1), there are solid-state alternatives to ion chambers. Photodiodes may be employed, and avalanche photodiodes (APDs) for fast detection (ns). Neither of these is energy selective, and thus may also require a filter-slit assembly to improve the signal/noise ratio. Neither has the advantage of the controllable intrinsic sensitivity control of an ion chamber, through (partial) gas pressures. Also, each can suffer from crystal glitches within the detector.

Energy discriminating solid-state systems are generally the detectors of choice. The detector diode element may be based on silicon (e.g., in a drift diode), or on germanium (e.g., a *pin* diode). The latter is a more absorbing material, an advantage at higher x-ray energies. These are more complex pieces of instrumentation, with arrays of diodes gathered in a detector head to

Table 4.2 Energies (keV) of edges and emissions of the L_3 absorption of elements of the third long period and potential filters for fluorescence measurements. Gray shading indicates degree of problems monitoring with the emission line using the suggested filter.

Element	L_3	$L\alpha_1$	$L\beta_1$	Filter	K/L_3	$K\alpha_1/L\alpha_1$	$K\beta_1/L\beta_1$
Barium	5.25	4.66	4.83	Titanium	4.97	4.51	4.93
Lanthanum	5.48	4.65	5.04	Titanium	4.97	4.51	4.93
Hafnium	9.56	7.90	9.02	Copper	8.98	8.05	8.91
Tantalum	9.88	8.15	9.34	Copper	8.98	8.05	8.91
Tungsten	10.21	8.40	9.67	Tantalum	9.88	8.15	9.34
Rhenium	10.54	8.65	10.01	Tungsten	10.21	8.40	9.67
Osmium	10.87	8.91	10.36	Tungsten	10.21	8.40	9.67
Iridium	11.22	9.18	10.71	Germanium	11.10	9.89	10.98
Platinum	11.56	9.44	11.07	Germanium	11.10	9.89	10.98
Gold	11.92	9.71	11.44	Germanium	11.10	9.89	10.98
Mercury	12.28	9.99	11.82	Gold	11.92	9.71	11.44
Thallium	12.66	10.27	12.21	Gold	11.92	9.71	11.44
Lead	13.04	10.55	12.61	Gold	11.92	9.71	11.44
Bismuth	13.42	10.84	13.02	Lead	13.04	10.55	12.61
Polonium	13.81	11.13	13.48	Bismuth	13.42	10.84	13.02
Astatine	14.21	11.43	13.88	Bismuth	13.42	10.84	13.02
Radon	14.62	11.73	14.32	Bismuth	13.42	10.84	13.02

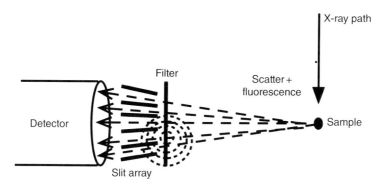

Figure 4.20 Schematic of a filter and Soller slit assembly in front of a fluorescence detector.

Figure 4.21 64-element germanium detector on I20 (*Source:* Diamond).

increase the solid angle that is collected (Figure 4.21). Both require the detector head to be cooled during use. For silicon-based systems cooling can be afforded by a thermoelectric device, but liquid nitrogen temperatures are essential for the germanium.

Returning to the example of the nickel spectrum in Figure 4.16, we can see that the $K\alpha$ energy for Ni is ~500 eV to lower energy of a typical XAFS scan (Ni K edge 8333 eV) (Figure 4.22). Energy discriminating detectors can afford resolutions in the range of ~130–300 eV, and thus can be used to isolate the signal due to the $K\alpha_1$ and $K\alpha_2$ lines from the background elastic scatter, and the $K\beta$ emission.

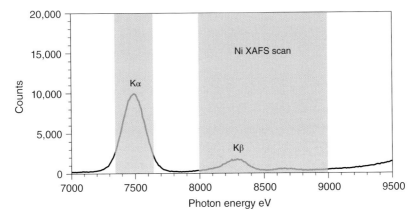

Figure 4.22 Fluorescence and scatter signals from a Ni-containing sample with a multi-channel analyzer (MCA) highlighting the energy window of a XAFS scan and also of selective *Kα* monitoring (*Source:* Data from I20, Diamond).

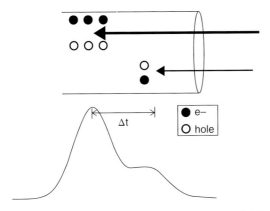

Figure 4.23 Events related to two x-ray photons of different energy impacting on a diode in an energy discriminating detector.

The principle behind the detector is illustrated in Figure 4.23. The absorption of an x-ray photon resulting in a cascade of electron hole pairs in the diode. The number of pairs is related to the x-ray energy (Section 4.3.1) and thus forms a measure of that energy. Thus the magnitude of the Gaussian curves in the figure relate to the measurement of these events using a multi-channel analyzer (MCA). The resolution of the time difference relates to the count rate that can be accommodated by the detector before saturation. This is limited by the peaking time of the device. Silicon drift devices can be obtained with a peaking time of 0.1 μs, giving a count rate of 900 kHz. This high count-rate is obtained at the expense of energy resolution; the precision of the electron-pair count deteriorates with a shorter acquisition time. For an authentic XAFS spectrum, it is wise to keep below the saturation limit to ensure a linear response, and a count rate of 100 kHz per diode is generally safe.

Under ideal conditions, ratio of the fluorescence signal, I_f, to the incoming radiation, I_0, is proportional to the x-ray absorption (equation 4.11). And so it can be expected that this is a straightforward means of measuring a XAFS spectrum. However, this only holds for particular conditions: dilute and/or thin. In these conditions, for example, a complex of a heavy element, for example, silver dissolved in a light-atom matrix, for example, water, the absorption due to the matrix is essentially constant across the spectrum (Ag K edge 25.5 keV) and the thickness sampled is unchanged. The sample would be oriented to maximize I_f, with the sample set at 45° to the incoming beam ($\theta = \phi = 45°$ in Figure 4.24).

$$I_f / I_0 \propto \mu(E) \tag{4.11}$$

However, if the sample is significantly absorbing, then the penetration depth will increase with energy above the edge. The fluorescence photons, for example, $K\alpha$ are of an energy close to the low point of the mass absorption of an element just before the edge. If the sample is concentrated then the fluorescence yield will increase significantly with photon energy. This will make the background subtraction more difficult but it also has the effect of dampening the EXAFS features so reducing the apparent coordination number. One way to reduce these sample absorption effects is to rotate the sample close to normal to the incoming beam such that $\phi \sim 0$ (grazing exit). That will strongly reduce the signal reaching the detector, but it will remove the energy dependence giving the relationship in equation 4.12, where $\mu_{total}(E_f)$ is the mass absorption of the entire sample at the energy of the fluorescence.

$$I_f / I_0 \propto \frac{\mu(E)}{\mu_{total}(E_f)/\sin\phi} \tag{4.12}$$

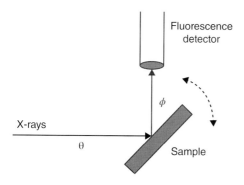

Figure 4.24 Sample orientation variation in a fluorescence measurement.

Alternatively, if the sample is thin, and observed at grazing incidence ($\theta \sim 0$), then the I_f/I_0 is proportional to sample x-ray absorption and the thickness (equation 4.13). Grazing incident angle XAFS is an established technique for investigating thin films, often in tandem with x-ray reflectivity measurements. The maximum sensitivity is obtained if θ is just under the critical angle.[9]

$$I_f / I_0 \propto \mu(E)t \tag{4.13}$$

4.3.3.2 High-Resolution Fluorescence Detection (HERFD) and X-Ray Emission Spectroscopy (XES)

In the previous section, we started with the detection of the total x-ray fluorescence yield, and with energy discriminating detectors arrived at the capability of windowing onto a set of closely spaced transitions, for example, $K\alpha_1$ plus $K\alpha_2$. Improving the energy resolution further will allow the refinement of investigating these two channels independently. To achieve that resolution the emission spectrum is discriminated using a crystal analyzer in a secondary spectrometer (Figure 4.25). The analyzer crystal is bent to the radius of curvature of the Rowland circle that also includes the sample and detector.[10] The circle may be close to the horizontal or vertical planes and may also be stacked with a series of analyzer crystals to increase the solid angle of collection and thus the sensitivity.[11,12] In some installations a small position sensitive detector is used as it eases greatly the alignment of the optical system.

The resolution is achieved using a very high Bragg angle, and benefits from a small source size. Energy scanning in the secondary spectrometer is achieved by rotating the monochromators onto a new position on a Rowland circle and

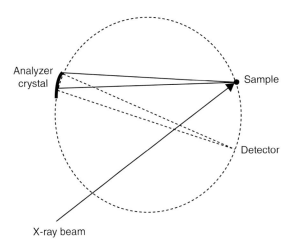

Figure 4.25 Johann geometry for energy resolution of the emission from a sample using a point detector.

translating the detector to keep it on the new circle. A suite of crystals is required to cover different emission lines (Table 4.3). The path length of the fluorescence signal is considerably increased, depending upon the radius of the Rowland circle. To avoid strong signal loss to the absorption by air, these spectrometers are generally mounted within a helium-filled shroud. Such an arrangement is shown in Figure 4.26.

The x-ray emission spectra of copper metal and its oxides (Figure 4.27) show little structure, with only a small shift (< 1 eV) to higher energy. Such scans can provide the optimum emission energy to maximize the sensitivity of fluorescence detection for each sample. An important observation was that the XANES features that were observable using this partial fluorescence

Table 4.3 Examples of crystal reflections and diffraction angles required the measure the x-ray emission spectra of some emissions of *3d* elements (as for I20, Diamond).

Element	$K\alpha_1$ (eV)	Reflection, θ°	$K\beta_1$ (eV)	Reflection, θ°
Ti	4511		4932	Si(400), 67.805
V	4952	Si(400), 67.217	5427	Ge(422) 81.525
Cr	5415	Ge(422), 82.424	5947	Si(333), 85.812
Mn	5899	Si(422), 71.403	6490	Si(440), 84.199
Fe	6404	Ge(440), 75.398	7058	Si(531), 73.038
Co	6930	Si(531), 77.018	7649	Ge(444), 82.887
Ni	7478	Si(620), 74.881	8265	Si(444), 73.104
Cu	8048	Si(444), 79.342	8905	Si(642), 73.58
Zn	8639	Si(642), 81.397	9572	Ge(660), 76.288

Figure 4.26 View over mount for three-analyzer crystals through the helium shroud to the sample position (*Source:* Diamond, I20).

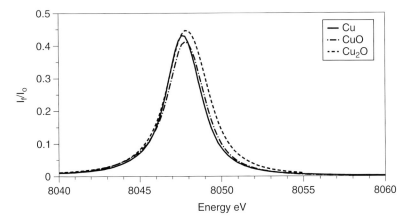

Figure 4.27 $K\alpha_1$ x-ray emission spectra of Cu, Cu_2O, and CuO recorded with a four-bounce Si(111) monochromator for I_0 and three Si(444) XES analyzer crystals (*Source:* Diamond, I20).

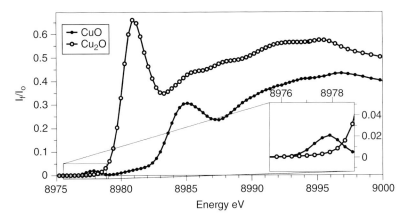

Figure 4.28 K edge HERFD of copper oxides detected using the $K\alpha_1$ emission line of copper (*Source:* Diamond, I20).

yield method of detection, which has a much higher resolution that the width of the emission line, were considerably sharper than those recoded by transmission or by total fluorescence yield,[13] and the technique is now termed high-energy resolution fluorescence detection (HERFD). In the XANES region, the lifetime broadening is dominated by the core-hole after the emission. In the case of $K\alpha_1$ emission, the most intense, that will be the L_3 level. The core-hole line-widths for Cu 1.55 and 0.56 eV for the K and L_3 shells, respectively. The effects are illustrated in Figure 4.28 for the two copper oxides. The sharpness of the edge features makes discrimination between the two oxidation states very clear. Also, by reducing the background slope of the intense features to low energy, weak pre-edge features (for CuO) can be identified more obviously. Above the XANES region line-width sharpening is lost and there is rarely a gain in recording EXAFS in this way rather than by an energy-selective detector that will generally have a larger solid angle and fluorescence signal.

Other emission lines can also be monitored. In the *K* series, as is illustrated in Figure 4.29 for copper, the relative intensity follows an order:

$$K\alpha_1, KL_3 (100) > K\alpha_2, KL_2 (\sim 50) > K\beta_1, KM_3 (\sim 20) > K\beta_3, KM_2 (\sim 10)$$

It is evident that the line-width of the $K\beta_{1,3}$ emission is rather wider than that of the two *Kα* lines. A comparison of the HERFD of copper foil measured with the $K\alpha_1$ and $K\beta_{1,3}$ emissions is shown in Figure 4.30. The resolution also appears

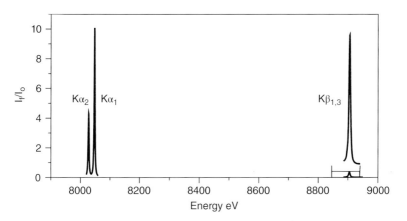

Figure 4.29 *K* emission lines of copper metal (*Source:* Diamond, I20, data from Roberto Boada-Romero).

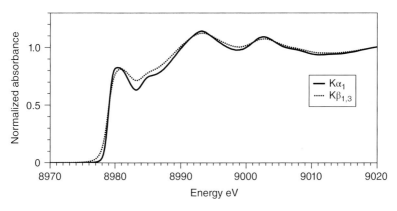

Figure 4.30 Cu *K* edge HERFD of copper foil measured with emissions from *Kα₁* (8047.8 eV, 4 scans) and *Kβ₁,₃* (8905.3 eV, 12 scans) (Source: Diamond, I20).

to be better for the former, more intense feature. For these spectra, the widths of the elastic peak were ~1.2 and ~1.7 eV, respectively, for $K\alpha_1$ and $K\beta_{1,3}$. One factor for this is the different resolutions achievable for these lines with the emission spectrometer. The $K\alpha_1$ line was measured using Si(444) reflections at 79.3174°, which ideally gives a resolution of 0.026 eV/mdeg; the lower Bragg angle employed for the $K\beta_{1,3}$ emission (73.5809°) results in a lower resolution (0.047 eV/mdeg) in spite of the higher order reflection (642). However, the core-hole lifetimes will also contribute.[14,15] The natural widths (Γ) of the fluorescence lines are closely related to those of the two states involved.

$$\Gamma(K\alpha_1) = \Gamma(K) + \Gamma(L_3) \tag{4.14}$$

$$\Gamma(K\beta_{1,3}) = \Gamma(K) + \Gamma(M_{2,3}) \tag{4.15}$$

When the exciting energy is close to threshold (E_o), the emission broadening is reduced[13,16] and can approach a FWHM of 2 $\Gamma(L_3)$ for the $K\alpha_1$ emission (and 2 $\Gamma(M_{2,3})$ for $K\beta_{1,3}$). The natural line-widths of the K and L_3 edges of copper are 1.55 and 0.56 eV, respectively. A sharpening of the XES line to ~1.2 eV near the absorption edge is attainable, and thus affords the improved resolution of the HERFD spectrum (Figure 4.28). However, the widths due to lifetimes of the core-holes of the $M_{2/3}$ edges (1.6 eV) are very similar to that of the K edge, and in this case $K\beta_{1,3}$ HEFRD does not improve the resolution of the XANES region over a conventional transmission spectrum.

4.3.3.3 Resonant Inelastic X-Ray Scattering or Spectroscopy (RIXS)

The capability of scanning both incoming and outgoing radiation opens up new possibilities. As can be seen for CuO (Figure 4.31), the shape of the emission spectrum varies with excitation energy. Using the excitation energy that affords the maximum emission (8995 eV), a single line is observed. Lowering the excitation energy not only reduces the total emission, but it also changes the shape of the emission curve. Mapping the variations in emission with both excitation and emission energy are experimentally feasible by stepping though those energies on the primary and secondary crystal spectrometers, respectively. The technique is variously described as RIXS (Resonant Inelastic X-Ray Scattering or Spectroscopy) or RXES (Resonant X-Ray Emission Spectroscopy).[17,18]

The measurement of RIXS data is achieved by recording a series of XAFS spectra with a stack of emission energies, and the natural way to present this is as in Figure 4.32a. Here there is a contour map of the I_f/I_0 values for the energy of incoming and outgoing x-radiation. The horizontal line through the emission maxima (Figure 4.31) is the HERFD spectrum in Figure 4.28. The vertical lines will be emission spectra measured at different excitation energies. The variations shown in Figure 4.31 can be envisaged as coming from different slices through the structured patterns below an excitation energy of ~8990 eV.

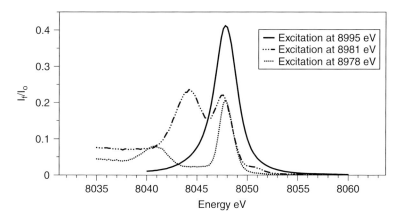

Figure 4.31 *Kα₁* emission spectra of CuO with three excitation energies. The XES with the two lower energy excitations are amplified by a factor of 10 (*Source:* Diamond, I20).

There is an alternative presentation intrinsic to the inelastic scattering associated with the overall process. In that the *y*-axis takes the form of the difference between the excitation and emission energies, which is the energy transferred to the sample as it is converted into an excited sate with a *2p* core hole (Figure 4.32b). In this representation the HERFD spectrum is the diagonal through the maxima. In both representations of RIXS, the pre-edge peak evident in the HERFD spectrum can be seen to be a resonance to low energy of the absorption edge along the line of constant emission energy (CEE). Here vertical lines represent emission spectra at constant incident energy (CIE) and the horizontal lines XAFS spectra at constant emission energy (CEE).[17,18]

The RIXS patterns are very sensitive to oxidation state and coordination number. Comparison between the plots in Figure 4.33 show very distinct differences for the linear Cu^I site in Cu$_2$O and the tetrahedral [Cu(dmp)$_2$]$^+$ near 8992 eV. For the oxide the pre-edge feature is very intense and to relatively low energy. In the linear geometry two of the *4p* orbitals are non-bonding and thus provide the route to a relatively low energy dipole allowed transition. In the tetrahedral example these orbitals will largely have copper-nitrogen antibonding character, and so be raised in energy. This complex also showed a small pre-edge resonance, about 2 eV higher in energy than that observed for CuO (Figure 4.31).

4.3.3.4 Inelastic X-Ray Raman Scattering (XRS) or Nonresonant Inelastic X-Ray Scattering (NIXS)

The range of inelastic scattering process that can be monitored with secondary spectrometers is large. It is very dependent upon the resolution of the monochromator, so electronic and vibrational processes can be identified with sufficiently high resolution (sub eV to meV). With a resolution typical of the

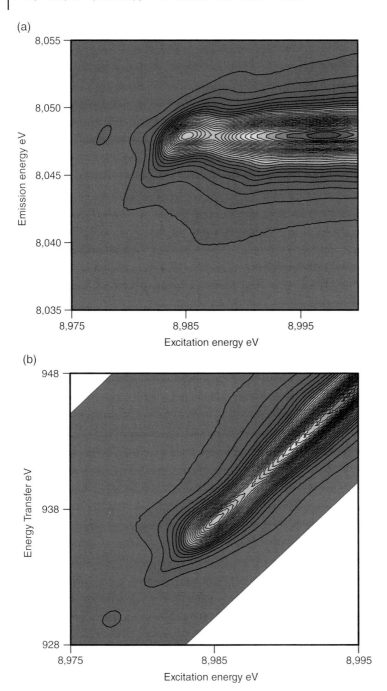

Figure 4.32 *Kα₁* RIXS spectra of CuO plotted with 32 contours of I_f/I_0. Axes used are the excitation and a) emission or b) transfer energies (*Source:* Diamond, I20).

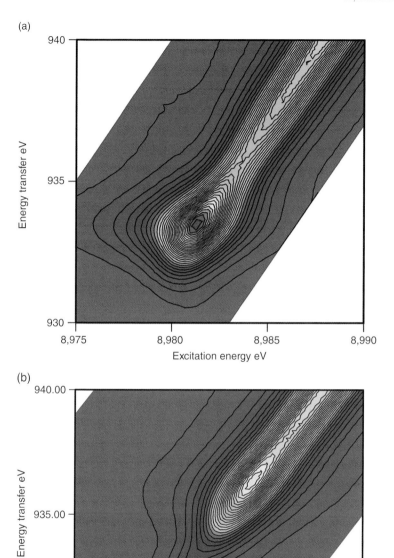

(a)

(b)

Figure 4.33 *Kα₁* RIXS spectra with 32 contours of I_f/I_0 of two Cu(I) samples. a) Cu_2O b) $[Cu(dpm)_2]PF_6$ (dpm = 2,9-dimethyl-1,10-phenanthroline) (*Source:* Diamond, I20).

XES and RIXS measurements described above, then features with an energy transfer of a few 10s and 100s of eV have been identified as the absorption edges of low Z-elements in the sample.[19,20] A demonstration example is provided by the observation of the carbon K edges of graphite and diamond (Figure 4.34). This was using excitation energies of 8400 or 8900 eV. Clearly sensitivity is challenging, but installations are in place with multiple analyzer-detector arrays to enhance the solid angle that can be collected for a particular energy loss at different scattering angles [21]. As a result, it has become feasible to investigate the structures of light elements within bulk materials, such as the study of lithium and oxygen in discharged electrodes.[22]

4.3.4 X-Ray Excited Optical Luminescence (XEOL)

In addition to stimulating x-ray fluorescence, x-ray absorption can, for some samples, give rise to a detectable emission in the uv-visible region. The arrangement for the detection of such emissions is very similar to that for x-ray fluorescence.[23] The fluorescence can be collected via a lens and then passed into a box shielding the detector from x-radiation (Figure 4.35). The XEOL effect can be used as a measure of x-ray absorption, selecting the energy of interest

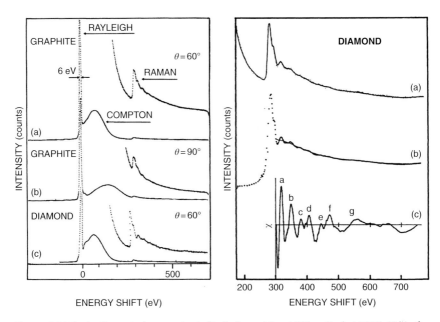

Figure 4.34 Inelastic scattering spectra Left: a) of graphite at 60° excited at 8900 eV, b) of graphite at 60° excited at 8900 eV, c) of diamond at 60° excited at 8400 eV. Right: Extraction of the C K edge EXAFS of diamond (*Source:* Tohji 1989.[20] Reproduced with permission of American Physical Society).

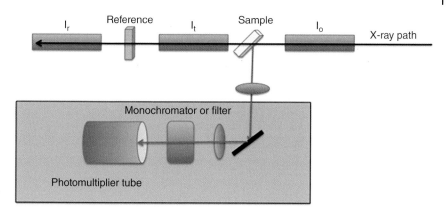

Figure 4.35 Schematic of instrumentation for carrying out x-ray excited optical luminescence (XEOL).

with an optical filter, and then passing through to a photomultiplier tube or photodiode; this is analogous to having an energy selective x-ray detector. Alternatively, using an optical monochromator, element-selective optical emission spectra can be recorded. In addition, the emission may be due only a portion of the sites adopted by the element being studied, and so the method will afford site selective XAFS. An example is the selection between recording the XANES of ZnO or ZnS by selecting appropriate emission wavelength (510 and 575 nm, respectively).[24]

Such instrumentation[25] was employed to probe the nature of nanocrystal-line germanium prepared by laser ablation and other methods.[26] The wavelength of the optical emission was very dependent upon the mode of sample preparation. Laser ablation providing an XEOL signal around ~700 nm that could be using to record optically detected (OD) XAFS spectra (Figure 4.36). Although the conventionally measured XAFS spectrum indicated elemental germanium predominated in the sample, the source of the emission could be ascribed to GeO_2. Time-resolved XEOL effects can also be measured,[25] as illustrated by the recording of Eu L_3 edge XANES of Eu_2O_3 from both short-lived (360 nm) and long-lived (610 nm) states.

4.4 Spatial Resolution

4.4.1 Methods of Studying Textured Materials

The most common methods of investigating the composition of materials with a resolution of microns or nm are scanning and transmission electron micros-copy. At one level this provides a measure of the electron density of voxels in a material, but with the addition of an x-ray fluorescence analyzer, the maps can

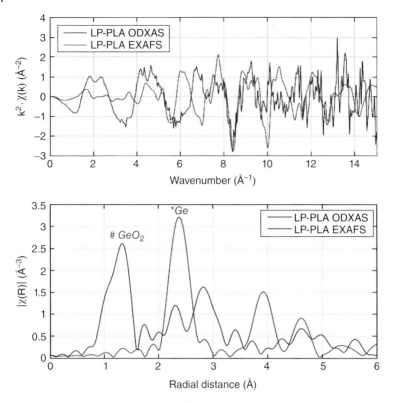

Figure 4.36 Comparison of the optically detected (OD) (700 nm emission) and transmission Ge *K* edge EXAFS (top) and Fourier transform (bottom) of nanocrystalline germanium prepared by laser ablation (LP-PLA) (*Source:* Courtesy of Andrei Sapelkin, Queen Mary University).

show elemental composition. In principle, the techniques described so far can be adapted to mapping of sites within a textured material, although in practice there are limitations, most commonly of sensitivity. Chemical or site speciation could be envisaged by variation in XANES, XMCD, or XEOL properties. The two main methods are *full field*, in which the spatial resolution is derived after the sample, and *scanning-transmission* where a highly focused beam is tracked across the sample.

4.4.2 Full-Field Transmission X-Ray Microscopy (TXM)

A schematic of a full field transmission x-ray microscope (TXM) is shown in Figure 4.37. The source used is a bending magnet. Generally this has been preferred over an undulator for this approach. The smaller beam size of an undulator increases the space over which the x-ray beam has high degree of

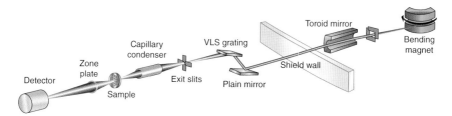

Figure 4.37 Schematic of a soft x-ray full-field transmission x-ray microscope (*Source:* Diamond, B24).

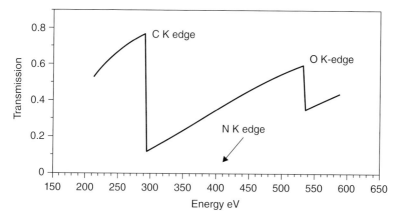

Figure 4.38 Transmission calculated for a 1 μm sample of ethanol in the C, N, and O *K* edge regions.

coherent character. This can result in interference fringes from any element in the x-ray path and thus can introduce artifacts into the image of the sample itself. For this beamline a capillary system is used to focus the beam size allowing illumination of the sample over a 20-μm range. The transmitted beam is refocused by a zone-plate onto an imaging detector providing a spatial resolution of ~35 nm. The image is acquired using a monochromatic beam at a single energy, here with a resolving power of 1000. An important use for such a microscope is for biological samples in an aqueous medium. In that case, the using the "water window" provides considerable advantage. The attenuation length of water rises from 0.8 μm near 200 eV to 10 μm just below the *K* absorption edge of oxygen. The x-ray transmission of an organic sample is modeled in Figure 4.38 by 1 μm of ethanol. Near 500 eV the reduction in transmission caused by this is ~50%, and thus there is considerable scope for absorption contrast with a relatively transparent background. Samples can be cryo-cooled, which enhances the stability under x-radiation. The samples may also be tilted

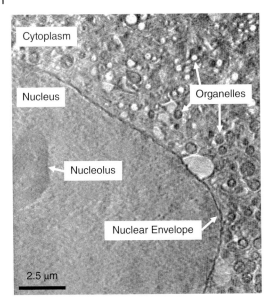

Figure 4.39 Slice of a three-dimensional construct of a neuron-like mammalian cell taken by cryogenic full field transmission x-ray microscopy using 500 eV radiation (*Source: Courtesy of Elizabeth Duke and Michele Darrow).*

to generate a series of images. A three-dimensional tomograph of the sample can be generated from the tilt-series.[27]

An example of application is the imaging of a neuron-like mammalian cell line (PC-12) expressing huntingtin exon fragments (Figure 4.39) The huntingtin gene is associated with Huntington's disease; this sample was prepared by Wei Dai, in the lab of Wah Chiu, at Baylor College of Medicine, Houston, Texas, and imaged by Michele C. Darrow, in the laboratory of Elizabeth M.H. Duke, at Diamond. The beamline in question operates with x-ray energy of 500 eV with a stable sampling arrangement to allow for rapid image acquisition. The figure shows a slice of a reconstructed three-dimensional volume from absorption contrast images collected by cryogenic tomography. The image clearly shows the major components of the cell. Some of the organelles (mitochondria, lipid droplets, and autophagosomes) can be identified by their characteristic features. The large organelle near to the nuclear membrane was of particular interest.

This technique can be used at higher x-ray energies, and this will allow less thin samples to be used for inorganic materials. Elemental maps can be obtained by comparing images before and after an absorption edge, and XANES spectra derived by careful treatment of images recorded with different x-ray energies near an absorption edge of interest.[28] Care must be taken due to the chromatic nature of the diffraction imaging of a zone plate and also the change in background absorption of the sample.

4.4.3 X-Ray Photoelectron Emission Microscopy (X-PEEM)

A very different full-field approach involves use of an electron microscope to capture images from photoelectrons emitted the sample surface under the x-ray beam. Typically this is employed in the soft x-ray region (< 2000 eV), a fertile region for studying magnetic properties using the L edges of $3d$ elements. The ultra-high vacuum conditions preferred of optimizing the electron microscope are consonant with the high absorption coefficients of samples in this energy regime and also on the need for surface cleanliness for what is a surface-selective technique. Spatial resolutions of 20–30 nm can be achieved. By using XMCD methods, magnetic domains can be imaged using this method, as illustrated for a film of $La_{0.7}Ca_{0.3}MnO_3$ on $BaTiO_3$ (Figure 4.40). This material

(a)

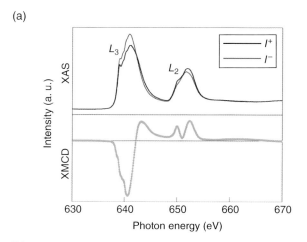

(b)

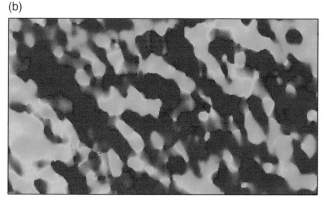

Figure 4.40 a) XAFS with right (I^+) and left (I^-) circular polarization and electron yield detection and the resulting XMCD signal of the Mn $L_{2/3}$ edges of $La_{0.7}Ca_{0.3}MnO_3$ on $BaTiO_3$ at 210K and an applied magnetic field of 0.5. b) A zero field Mn PEEM image at ~150 K; image resolution 50 nm (*Source:* Courtesy of Sarnjeet Dhesi).

displays a giant magnetocalorific effect close to temperature (~200 K) of a phase change of the $BaTiO_3$ substrate.[29] The image is based on the Mn XMCD contrast measured using the Mn L_3 (~640 eV) and L_2 (~650 V) edges, and shows different orientations of magnetic moments of Mn in the domains.

4.4.4 Focused-Beam Microscopies

The alternative approach to imaging is to raster the sample against a highly focused x-ray beam. In principle, the effects of the x-rays on the sample can be measured by all the XAFS techniques described in Section 4.3, and also by x-ray diffraction.

4.4.4.1 Scanning Micro- and Nano-Focus Microscopy

An example schematic for a microfocus spectroscopy beamline, I18 at Diamond, is shown in Figure 3.4. Medium energy monochromatic light is focused onto the secondary slits, which act as the source for the second stage of focusing using Kirkpatrick-Baez (KB) mirrors (Section 3.3.3.1). The result is a beam size of the order of 1 μm. An elemental map of the sample can be built up using the output of the MCA of an energy-selective fluorescence detector to estimate the elemental content at each pixel. These measurements are stored with the array of positional data as the x-ray beam is rastered over the sample; a spatial map can be constructed from these files. Oxidation state maps may be derived from measurements at particular energies within the XANES region. Additionally, XAFS spectra may be recorded at regions of importance within the sample. An example of how this is used to speciate a textured sample is shown in Figure 4.41. The sample is a sliver of wood (Figure 4.41a) taken from the Henry VIII's warship, the *Mary Rose*. The ship sank in 1545 after being in service for 34 years, and was raised from the bed of the Solent in 1982. Since then there has been a major conservation project emanating from the museum in Portsmouth. The sulfur fluorescence maps (Figure 4.41b) show a low proportion of reduced sulfur, with the majority of the element being in the form of sulfate(VI). The maps of Fe fluorescence display a very small proportion of Fe(II), with the majority of the iron being present as Fe(III) (Figure 4.41c). The elements occupy very similar regions of the sample, and it is considered that iron is associated the oxidation of organic sulfur to sulfate. This results in highly corrosive localities, and conservation is designed to mitigate this.[30]

The spatial resolution can be enhanced by increasing the ratio of the distances between the secondary source and the focusing elements and the focusing element to the sample. In practice this requires a long beamline, as in I14 of the Diamond (Figure 4.42). Substitution of the KB mirrors by a zone plate will further reduce the beam size to the order of 10 nm, but this probably restricts the viable energy range of XAFS scans (see Section 3.3.3.3).

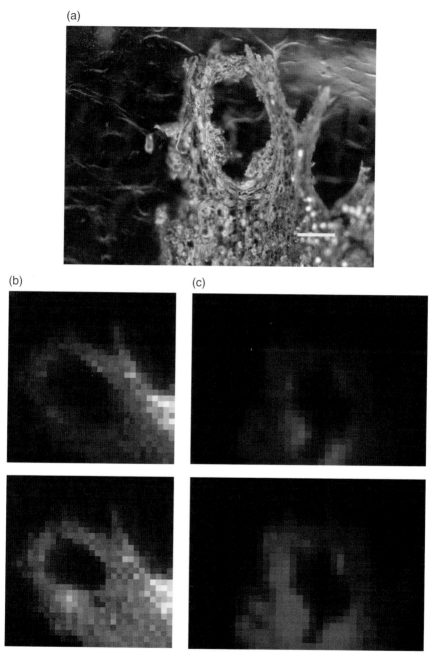

Figure 4.41 Images of a sliver of wood from the Tudor warship *Mary Rose*. a) Optical image. b) Sulfur fluorescence map with excitation at 2473.1 eV (upper) and above 2482.1 eV (lower). c) Fe fluorescence map of Fe(II) (upper, amplified 4X) and total Fe (lower) (*Source:* Courtesy of Andy Smith, University of Manchester).

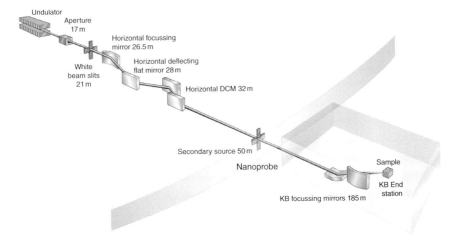

Figure 4.42 Schematic of the initial layout for beamline I14 at Diamond.

4.4.4.2 Scanning (Transmission) X-Ray Microscopy (STXM)

A scanning transmission x-ray schematic is presented in Figure 4.43, for beamline I08 at Diamond. Typically for a STXM, this beamline is based upon an undulator source, with a high degree of coherence. The coherence is required to give optimum resolution from the focusing zone plate.[31] The sample stage is rastered against the x-ray beam (250–4200 eV) and an absorption contrast image can be recorded on the CCD camera with a resolution of ~20 nm.

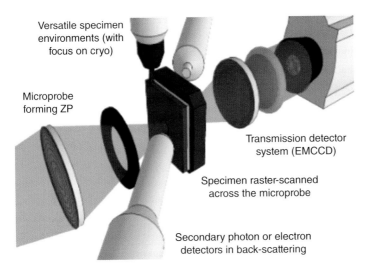

Figure 4.43 Schematic of the sample area of a scanning x-ray microscope (*Source:* Diamond, I08).

(a)   (b)

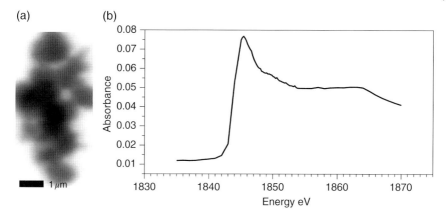

Figure 4.44 a) Absorption contrast image of a sol-gel product from the reaction of Si(NHMe)$_4$ with NH$_3$ observed above the silicon *K*-edge (1850 eV). b) Si *K* edge XANES of the imaged particle (*Source:* Courtesy of Andrew Hector, University of Southampton).

Phase-contrast imaging may also be performed. At each point in the sample position a XANES spectrum can be measured, and a map of constituents in the sample can be constructed post-acquisition using XANES features. As for the medium energy micro- and nano-focus beamlines, an energy selective fluorescence detector allows for element mapping. This particular beamline has a helical undulator as it source, and thus offers an alternative technique to X-PEEM for investigating magnetic materials.

An illustration of a materials application is the structural characterization of a sol-gel derived silicon nitride sample with high surface area[32] (Figure 4.44). Micron-sized particles are evident, which are uniform in terms of the silicon environment. The Si XANES spectrum is consistent with an environment like Si$_3$N$_4$ with negligible hydrolysis to silica.

4.5   Combining Techniques

Many secondary techniques, such as uv-visible, infra-red, and Raman spectroscopies can be incorporated into the sampling area, providing that care is taken to match the sampling criteria for both that and the XAFS measurements. These are varied and some examples are provided in Chapter 6. In this chapter we concentrate on the main techniques also probed by the x-radiation of the beamline.

4.5.1   Two-Color XAFS

One aspect of this is to investigate a sample by acquiring spectra from more than one absorption edge. Monochromator and mirror settings can be stored and routines scripted to allow a sequence of edges to be studied. This can be achieved rapidly providing there are no changes in mirror or monochromator. If the sample is stable enough to allow this change then this is entirely viable. The sample concentration or thickness will not be optimized for each edge, but a compromise may well be feasible.

A slightly more sophisticated approach is to carry out a double-edge study using the energy dispersive method (Section 4.2.2) with sufficiently large band spread as to span the XAFS range of two edges. Unlike the previous example, this type of measurement is genuinely synchronous and thus can be used for unstable systems and chemical change. An early example was the monitoring of the formation of platinum-germanium catalyst particles from organometallic precursors;[33] the Pt L_3 (11564 eV) and Ge K (11103 eV) are close enough to make both accessible but separated enough for the contamination of the platinum-edge spectrum by that of germanium to be substantial. As the energies of the absorption edges are increased, the Bragg angle difference required to span a particular energy range is reduced. So it proved possible to obtain EXAFS for both the Rh (23220 eV) and Pd (24350 eV) K edges using a Si(111) polychromator[34] to during an *in situ* catalysis study of the reduction of NO.

Investigation of a sample using both hard and soft x-ray energies would need two independent optical paths suitable for each energy range. This could be achieved with a split take-off from a bending magnet, but a higher specification would by achieved with two different undulators with widely different characteristics. An example of this approach is Beamline I09 at Diamond (Figure 4.45). The line has the two undulators in an 8-m-long straight section, and these are canted so that the two beams may be separated into their own optical systems. These are recombined onto the sample within a UHV environment. A hemispherical electron energy analyzer enables hard x-ray photoelectron spectroscopy (HAXPES) and XAS spectra to be recorded by, for example, total electron yield (Section 4.3.2). This can be illustrated with application as part of a major study into safer lithium ion cathodes following lithium ion doping into the phase ε-LiVOPO$_4$.[35] The O K edge XANES was sensitive to the degree of additional lithium incorporation due to the accompanying reduction of vanadium affecting the V-O states.

4.5.2   X-Ray Scattering

A clear complementarity is achieved by incorporating x-ray diffraction into an experimental setup, as illustrated in Figure 4.46. The sample position is at the center of the circle containing the arc of the detector. Generally, XAFS spectra and powder diffraction patterns are taken successively, with the monochromator driven quickly to the energy of choice for the diffraction measurements. Providing the lifetime of the sample is at least of the order of minutes, the two techniques will be measuring the same sample. However, they may not equally

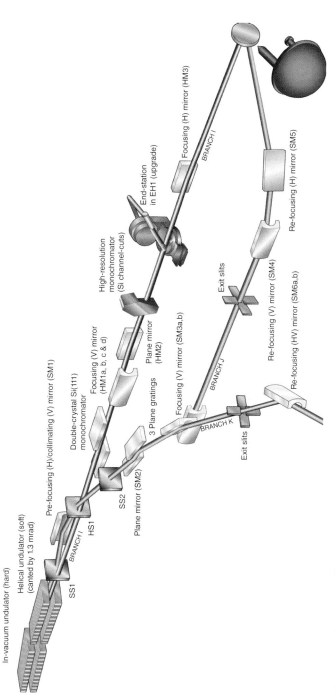

In-vacuum undulator (hard)

Helical undulator (soft)
(canted by 1.3 mrad)

Pre-focusing (H)/collimating (V) mirror (SM1)

SS1

BRANCH I

HS1

Double-crystal Si(111)
monochromator

Plane mirror (SM2)

SS2

Focusing (V) mirror
(HM1a, b, c & d)

3 Plane gratings

Plane mirror
(HM2)

Focusing (V) mirror (SM3a,b)

High-resolution
monochromator
(Si channel-cuts)

BRANCH K

Exit slits

Re-focusing (HV) mirror (SM6a,b)

Re-focusing (V) mirror (SM4)

BRANCH J

Exit slits

End-station
in EH1 (upgrade)

Focusing (H) mirror (HM3)

BRANCH I

Re-focusing (H) mirror (SM5)

Figure 4.45 Schematic of beamline I09 at Diamond.

Figure 4.46 Installation of a curved multi-element solid-state detector for wide-angle x-ray scattering on an XAFS beamline (Source: Diamond, B18).

probe the same components of a textured sample. The diffraction pattern will be dominated by components with a large mosaic providing a sharp, intense diffraction lines; the XAFS spectra will monitor the mean of the sites of the absorbing element, although for XAFS too, the sites of highest local order will provide the more evident backscattering. An optimal case for complementarity is a solid solution. The diffraction pattern will afford the long-range order with disordered occupancy of the lattice sites. XAFS measured for the component elements will reveal the distinct interatomic distances local to each element. In such circumstances both the diffraction and EXAFS results can be advantageously combined into a linked refinement procedure.[36]

4.6 X-Ray Free Electron Lasers (XFELs)

The number of stations equipped to carry out XAFS measurements at XFELs will increase significantly in 2017. Such experiments require their own sampling arrangements that are more stringent than those at synchrotron sources. They can extend the scope of XAFS studies into the fundamentals of processes.[37] In one sense, XFELs provide a level playing field. It is not the case that only tender, biological samples will suffer radiation damage; every sample must be considered tender and effectively one-shot sampling must be devised. Care of the effects on the exposed single-shot sample need also be taken. One of the benefits of having x-ray pulses of the order of 10 fs is that x-ray scattering takes place sufficiently fast that is can be measured before the sample undergoes a coulombic explosion; that is, when the sample area experiences a massive degree of ionization and the resultant charges repel and shoot away from each other. But the x-ray flux can be such that it may violate the model of a single photon exciting an atom and the resulting excited state being observed by XAFS methods.

4.6.1   Laser-Pump Measurements

The basic time sequence of a pump-probe experiment is shown in Figure 4.47. Initiation of the change is effected by a laser pulse, and the sample can be probed by the x-ray pulse after a possible delay period. In a synchrotron experiment this probe period can build up a time sequence of XAFS spectra, but this may be much more restricted in a FEL. There is a necessity to refresh the sample so that the next pulse interrogates the starting material. This pump-probe combination is set within cycles without the pump laser to derive a background. XAFS difference spectra between the light-on and light-off periods are generally analyzed. The excitation energy provided by the laser pump is most commonly in the uv-visible range. That is chosen to cause a specific photo-physical transition within the lifetime of the pulse (~20 fs). The structural and electronic changes within the sample are then probed by XAFS. There will be a cascade of secondary photo-physical and -chemical events dissipating this energy that may extend across several orders of magnitude of time units. In a solution or a solid matrix, this will end with local heating of the sample. Covering all of these events will involve experiments both at synchrotron and

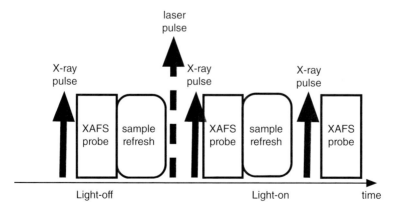

Figure 4.47   Schematic time sequence for a laser-pump, XAFS-probe experiment.

FEL sources. Alternative mechanisms for a pump source are i) to provide a heating pulse, for example into the electron sea of a metallic particle and ii) to excite a specific vibration within the sample in a multiphoton manner.

4.6.2 Sampling Environments

The intense energy sources of both pump and probe have consequences for the sampling methods. Our initial pump-probe solution experiments at synchrotron sources stumbled over several of these![38] We utilized highly unsymmetrical effective path lengths for transmission XAFS (~3 mm with pyrolytic carbon windows) and the laser excitation (sampled with a x-ray horizontal beam size of 5 µm). The recirculating solution was in contact with the surface of a quartz waveguide for the incoming laser. The build up of material on the surface of the quartz windows would result in increased energy absorption and fracturing of the ends of the waveguides. A subsequent experiment, also at ID24 of the ESRF, attempted to overcome this by recirculating the sample through a quartz capillary, which was translated during the experiment. Both the exciting laser and the polychromatic x-ray beam each caused deposition on the capillary and thus the spectra were contaminated by those of the decomposition product.

The preferred sampling arrangement for solutions avoids contact between the sample and the x-ray and laser windows.[39] The recirculating system is maintained but includes a liquid jet within the laser beam. The x-ray probe beam is positioned inside the excited portion of the jet. This is illustrated in Figure 4.48 in the installation at the SuperXAS beamline (SLS). The liquid jet can be seen inside an environmental chamber sitting in front of the fluorescence detection system consisting of a filter foil, Soller slits. and APD detector.

Figure 4.48 Liquid jet sampling system for time-resolved XAFS (SLS, X10DA).

For most other studies, windowless sampling has been chosen. Trains of protein crystals are delivered from a jet in a viscous liquid, and these may be mixed with reagent solutions via concentric glass capillaries to allow time-resolved crystallography of the products.[40] Similarly, a W L_3 XAFS study of WO_3 nanoparticles was carried out flowing a suspension.[41] Typical *in situ* heterogeneous catalysis studies within microreactors may have to be rethought. The suspension method may provide a route for liquid-solid systems or for those systems in which the active gaseous reagent is highly soluble. Gas-solid environments may possibly be investigated using thin or sputtered films on a sample belt.

4.6.3 X-Ray Beam Intensity

The high intensity can cause multiple ionizations of the absorbing atom, as has been reported for H_2S;[42] each sulfur atom absorbed an average of five photons. This situation is rather outside that of normal XAFS theory. This intensity can also result in non-linear effects like frequency-doubling, as has been reported for a germanium sample[43] using a highly focused beam to increase the flux density.

4.6.4 XAS and XES

But experimental conditions have been derived that are viable for XAFS studies.[44] A pump-probe study of the photophysics of [Ni(TMP)] (TMP = tetramesitylphorphyrin) utilized x-ray fluorescence detection at the Ni K edge to track the first ps of the cascade of processes;[45] a similar approach has been adopted to investigate [Fe(bipy)$_3$]$^{2+}$.[39] The instrument at LCLS also includes XES,[44] as has been used to investigate the manganese sites in photosystem II.[46] One of these systems is based upon the van Hamos geometry for a dispersive emission spectrometer (Figure 4.49).[47,48] This has the advantage over a Johann geometry in that an emission spectrum can be dispersed across the position sensitive detector (Figure 4.25). Thus one of the two dimensions of a RIXS plane is measured simultaneously, so dramatically reducing the number of measurements that need be made. This is of considerable benefit in time-resolved studies. There are complications however. As the Bragg angle changes across the image of the XES pattern, the resolution will also

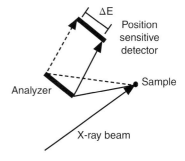

Figure 4.49 Schematic of a van Hamos geometry for an energy-dispersive x-ray emission spectrometer.

change across it. Also the solid angle collected for each energy point is very low, and so it is difficult to match the sensitivity and resolution of the Johann geometry. However, with a FEL source the pulse flux is extremely high and the van Hamos arrangement is proving to be very productive. [44,46]

4.7 Questions

1 One method of the neutralization of chlorine bleach, NaOCl, is to treat it with sodium thiosulfate, $Na_2S_2O_3$. You are assessing how to study this reaction *in situ* by XAFS since both chlorine and sulfur change oxidation states.

   A What are the attenuation lengths of an aqueous solution at the viable edges of these two elements?

   B What are the attenuation lengths of air at these energies?

   C What would be the transmission of a 25 micron film of a polyester window at these energies {Consider the polyester as $(-C_2H_4CO_2C_6H_4CO_2-)n$}?

   D Sketch a sampling system that could be successful for this experiment, giving your reasons.

2 Your project involves the synthesis of small particles of what you consider is Fe_3O_4. You synthesized this from a metallorganic, $[Fe(acac)_3]$, dissolved in ethanol, and then deposited the Fe_3O_4 onto a selection of supports, and are considering how to study the structure and properties of the iron.

   A The precursor solution has a concentration of Fe of 20 mM. Calculate the transmission of the sample before and after the Fe K edge for path lengths of 1 mm and 5 mm and the resulting edge jump. What path length would you propose to carry out a transmission measurement?

   B You have deposited the iron onto high area silica, SiO_2, at a concentration of 5 wt% and have made a pressed disc (density 1 gcm^{-3}) with a thickness of 200 μm. Calculate the transmission and edge jump of this sample. How would you propose to carry out this measurement? You are interested in the degree of crystallinity of this sample. How might you investigate that?

   C You have also deposited your sample onto the surface of a quartz crystal wafer. How might you investigate the structure and shape of the particles? How could you investigate their magnetic properties?

3 Another sample was prepared starting from $[Al(acac)_3]$, which you believe forms a colloidal material in aqueous ethanol. Subsequently you make similar supported samples.

   A How might you image the colloidal material in the aqueous medium?

   B You are interested in how the aluminum is distributed in the supported material. Design an experiment to investigate this.

4   You have a sample from a mining process and need to understand the envi-
    ronment of the minerals within the sample to optimize the next step in the
    process. The material has a clay base (aluminosilicate) and is believed to
    contain nickel, rhodium, silver, palladium, platinum, and gold. Upstream in
    the process had been a treatment with HCl. Your group is considering pur-
    chasing beam time on a line with an energy range of 2–12 keV fitted with
    an energy-resolving fluorescence detector based on silicon diodes with a
    multichannel analyzer with a quoted energy resolution of 150 eV at 5.9 keV.
    A   Assess the effectiveness of x-ray fluorescence mapping in examining
        the distribution of these elements across the sample?
    B   In this mixture which absorption edges would be used to characterize
        the materials by XANES?
    C   Which of these might be viable for an EXAFS analysis?

5   You are designing time-resolved experiments that will use an APD as a
    fluorescence detector and will need a filter before the detector to improve
    the signal/background ratio. By considering the energies of the absorption
    edges and main emissions,
    A   Propose a filter for an experiment at the vanadium K edge
    B   Consider the possibilities for studying the uranium L_3 edge.

References

1 'A new flexible monochromator setup for quick scanning x-ray absorption
   spectroscopy', J. Stötzel, D. Lützenkirchen-Hecht, R. Frahm, Rev. Sci. Instrum.,
   2010, **81**, 073109.
2 'Advances in high brilliance energy dispersive x-ray absorption spectroscopy',
   S. Pascarelli, O. Mathon, Phys. Chem. Chem. Phys., 2010, **12**, 5535–5546.
3 'Probing depth study of conversion electron/He ion yield XAFS spectroscopy
   on strontium titanate thin films', E. Yanase, I. Watanabe, M. Harada, M.
   Takahashi, Y. Dake, Y Hiroshima, Anal. Sci., 1999, **15**, 255–258.
4 'In situ studies of catalysts under reaction conditions by total electron yield
   XAS: possibilities and limitations of a new experimental technique',
   S. L. M. Schroeder, G. D. Moggridge, E. Chabala, R. M. Ormerod, T. Rayment,
   R. M. Lambert, Faraday Discuss., 1996, **105**, 317–336.
5 'Electron-yield x-ray absorption spectroscopy with gas microstrip detectors',
   T. Rayment, S. L. M. Schroeder, G. D. Moggridge, J. E. Bateman, G. E. Derbyshire,
   R. Stephenson, Rev. Sci. Instrum., 2000, **71**, 3640–3645.
6 'X-ray fluorescence yields, Auger, and Costner-Kronig transition probabilities',
   W. Bambynek, B. Crasemann, R. W. Fink, H.-U. Freund, H. Mark, C. D. Swift,
   R. E. Price, P. V. Rao, Rev. Modern Phys., 1972, **44**, 716–813.
7 'Atomic radiative and radiationless yields for K-shells and L-shells',
   M. O. Krause, J. Phys. Chem. Ref. Data, 1979, **8**, 307–327.

8 'An x-ray filter assembly for fluorescence EXAFS measurements', E. A. Stern, S. M. Heald, Nucl. Instrum. Meth. 1980, **172**, 397–399.

9 'Instrumentation for glancing angle x-ray absorption spectroscopy on the Synchrotron Radiation Source', S. Pizzini, K. J. Roberts, G. N. Greaves, N. Harris, P. Moore, E. Pantos, R. J. Oldman, Rev. Sci. Instrum., 1989, **60**, 2525–2528.

10 'A high-resolution x-ray fluorescence spectrometer for near-edge absorption studies', V. Stojanoff, K. Hämäläinen, D. P. Siddons, J. B. Hastings, L. E. Berman, S. Cramer, G. Smith, Rev. Sci. Instrum., 1992, **63**, 1125–1127.

11 'Five element Johann-type X-ray emission spectrometer with a single-photon-counting pixel detector', E. Kleymenov, J. A. van Bokhoven, C. David, P. Glatzel, M. Janousch, R. Alonso-Mori, M. Studer, M. Willimann, A. Bergamaschi, B. Henrich, M. Nachtegaal, Rev. Sci. Instrum., 2011, **82**, 065107.

12 'A seven-crystal Johann-type hard x-ray spectrometer at the Stanford Synchrotron Radiation Source', D. Sokaras, T.-C. Weng, D. Nordlund, R. Alonso-Mori, P. Velikov, D. Wenger, A. Garachtchenko, M. George, V. Borzenets, B. Johnson, T. Rabedeau, U. Bergmann, Rev. Sci. Instrum., 2013, **84**, 053102.

13 K. Hämäläinen, D. P. Siddons, J. B. Hastings, L. E. Berman, 'Elimination of the inner-shell lifetime broadening in x-ray absorption spectroscopy', Phys. Rev. Lett., 1991, **67**, 2850–2853.

14 'Natural widths of atomic *K* and *L* levels, *Kα* x-ray lines and several *KLL* Auger lines', M. O. Krause, J. H. Oliver, J. Phys. Chem. Ref. Data, 1979, **8**, 329–338.

15 'Appendix B. Core-hole lifetime broadening' in 'Unoccupied electronic states. Fundamentals of XANES, EELS, IPS and BIS', Ed. J. C. Fuggle, J. E. Inglesfield, Springer Verlag, Berlin, 1992.

16 'X-ray resonant Raman scattering: observation of characteristic radiation narrower than the lifetime width', P. Eisenberger, P. M. Platzman, H. Winick, Phys. Rev. Lett., 1976, **36**, 623–626.

17 'Core level spectroscopy of solids', F. de Groot, A. Kotani, CRC Press, 2008, Boca Raton.

18 'High resolution *1s* core hole x-ray spectroscopy in 3d transition metal complexes – electronic and structural information', P. Glatzel, U. Bergmann, Coord. Chem. Rev., 2005, **249**, 65–95.

19 'X-ray Raman scattering. Experiment I', T. Suzuki, J. Phys. Soc. Japan, 1967, **22**, 1139–1150.

20 'X-ray Raman scattering as a substitute for soft-x-ray extended-x-ray-absorption fine structure', K. Tohji, Y. Udagawa, Phys. Rev. B, 1989, **39**, 7590–7594.

21 a' Mulitelement spectrometer for efficient measurement of the momentum transfer dependence of inelastic x-ray scattering', T. T. Fister, G. T. Seidler, L. Wharton, A. R. Battle, T. B. Ellis, J. O. Cross, A. T. Macrander, W. T. Elam,

T. A. Tyson, Q. Qian, Rev. Sci. Instrum., 2006, **77**, 063901;b'A high resolution and large solid angle x-ray Raman spectroscopy end-station at the Stanford Synchrotron Radiation Lightsource', D. Sokaras, D. Nordlund, T.-C. Weng, R. Alonso Mori, P. Velikov, D. Wenger, A. Garachtchenko, M. George, V. Borzenets, B. Johnson, Q. Qian, T. Rabedeau, U. Bergmann, Rev. Sci. Instrum., 2012, **83**, 043112;c'A large-solid-angle x-ray Raman scattering spectrometer at ID20 of the European Synchrotron Radiation Facility', S. Huotari, Ch. J. Sahle, Ch. Henriquet, A. Al-Zein, K. Martel, L. Simonelli, R. Verbeni, H. Gonzalez, M.-C. Lagier, C. Ponchut, M. Moretti Sala, M. Krisch, G. Monaco, J. Synchrotron Radiat., 2017, **24**, 521–530.

22 'Bulk-sensitive characterization of the discharged products in Li-O₂ batteries by nonresonant inelastic x-ray scattering', N. K. Karan, M. Balasubramanian, T. T. Fister, A. K. Burrell, P. Du, J. Phys. Chem. C, 2012, **116**, 18132–18138.

23 'X-ray excited optical luminescence (XEOL) detection of x-ray absorption fine structure (XAFS)', L. Soderholm, G. K. Liu, M. R. Antonio, F. W. Lytle, J. Chem. Phys., 1998, **109**, 6745–6752.

24 'Synchrotron radiation x-ray excited optical luminescence for chemical state selective analysis', S. Hayakawa, T. Hirose, L. Yan, M. Morishita, H. Kuwano, Y. Gohshi, X-ray Spectrom., 1999, **28**. 515–518.

25 'A time resolved microfocus XEOL facility at the Diamond Light Source', J. F. W. Mosselmans, R. P. Taylor, P. D. Quinn. A. A. Finch, G. Cibin, D. Gianolio, A. V. Sapelkin, J. Phys. Conf. Ser., 2013, **425**, 182009.

26 'OD-XAS and EXAFS: Structure and luminescence in Ge quantum dots', A. Karatutlu, W. R. Little, A. V. Sapelkin, A. Dent, F. Mosselmans, G. Cibin, R. Taylor, J. Phys. Conf. Ser., 2013, **430**, 012026.

27 'Biological applications of cryo-soft x-ray tomography', E. Duke, K. Dent, M. Razi, L. M. Collinson, J. Microsc., 2014, **255**, 65–70.

28 'Three-dimensional imaging of chemical phase transformations at the nanoscale with full-field transmission x-ray microscopy', F. Meirer, J. Cabana, Y. Liu, A. Mehta, J. C. Andrews, P. Pianetta, J. Synchrotron Radiat., 2011, **18**, 773–781.

29 'Giant and reversible extrinsic magnetocaloric effects in La₀.₇Ca₀.₃MnO₃ films due to strain', X. Moya, L. E. Hueso, F. Maccherozzi, A. I. Tovstolytkin, D. I. Podyalovskii, C. Ducati, L. C. Phillips, M. Ghidini, O. Hovorka, A. Berger, M. E. Vickers, E. Defray, S. S. Dhesi, N. D. Mathur, Nature Mater., 2013, **12**, 52–58.

30 'The application of x-ray absorption spectroscopy in archaeological conservation: Example of an artifact from Henry VIII warship, the *Mary Rose*', A. V. Chadwick, A. Berko, E. J. Schofield, A. D. Smith, J. F. W. Mosselmans, A. M. Jones, G. Cibin, J. Non-Crystalline Solids, 2016, **451**, 49–55.

31 'Illumination for coherent soft x-ray applications: the new X1A beamline at the NSLS', B. Winn, H. Ade, C. Buckley, M. Feser, M. Howells, S. Hulbert, C. Jacobsen, K. Kaznacheyev, J. Kirz, A. Osanna, J. Maser, I. McNulty, J. Miao,

T. Oversluizen, S. Spector, B. Sullivan, Y. Wang, S. Wirick, H. Zhang, J. Synchrotron Radiat., 2000, **7**, 395–404.

32 'Sol-gel preparation of low oxygen content, high surface area silicon nitride and imidonitride materials', K. Sardar, R. Bounds, M. Caravetta, G. Cutts, J. S. J. Hargreaves, A. L. Hector, J. A. Hriljac, W. Levason, F. Wilson, Dalton Trans., 2016, **45**, 5765–5774.

33 'In situ, time resolved and simultaneous multi-edge determination of local order change during reduction of supported bimetallic (Pt-Ge) catalyst precursors using energy dispersive EXAFS', S. G. Fiddy, M. A. Newton, T. Campbell, J. M. Corker, A. J. Dent, I. Harvey, G. Salvini, S. Turin, J. Evans, Chem. Comm., 2001, 445–446.

34 'The impact of phase changes, alloying and segregation in supported RhPd catalysts during selective NO reduction by H_2', M. A. Newton, B. Jyoti, A. J. Dent, S. Diaz-Moreno, S. G. Fiddy, J. Evans, ChemPhysChem, 2004, **5**, 1056–1058.

35 'Uniform second Li ion intercalation in solid state ε-LiVOPO$_4$', L. W. Wangoh, S. Sallis, K. M. Wiaderek, Y.-C. Lin, B. Wen, N. F. Quackenbush, N. A. Chernova, J. Guo, L. Ma, T. Wu, T.-L. Lee, C. Schlueter, S. P. Ong, K. W. Chapman, M. S. Whittingham, L. F. J. Piper, Appl. Phys. Lett., 2016, **109**, 053904.

36 'Combined EXAFS and powder diffraction analysis', N. Binsted, M. J. Pack, M. T. Weller, J. Evans, J. Am. Chem. Soc., 1996, **118**, 11200–11210.

37 'Dynamic structure elucidation of chemical reactivity by laser pulses and X-ray probes', S. A. Bartlett, M. L. Hamilton, J. Evans, Dalton Trans., 2015, **44** 6313–6319.

38 'Energy dispersive XAFS: Characterization of electronically excited states of copper(I) complexes', M. Tromp, A. J. Dent. J. Headspith, T. L. Easun, X.-Z. Sun, M. W. George, O. Mathon, G. Smolentsev, M. L. Hamilton, J. Evans, J. Phys. Chem. B, 2013, **117**, 7381–7387.

39 'Femtosecond x-ray absorption spectroscopy at a hard X-ray free electron laser: application to spin crossover dynamics', H. T. Lemke, C. Bressler, L. X. Chen, D. M. Fritz, K. J. Gaffney, A. Galler, W. Gawelda, K. Haldrup, R. W. Hartsock, H. Ihee, J. Kim, K. H. Kim, J. H. Lee, M. M. Nielsen, A. B. Stickrath, W. Zhang, D. Zhu, M. Cammarata, J. Phys. Chem. A, 2013, **117**, 735–740.

40 'Mixing injector enables time-resolved crystallography with high hit rate at X-ray free electron lasers', G. D. Calvey, A. M. Katz, C. B. Schaffer, L. Pollack, Struct. Dynamics, 2016, **3**, 054301.

41 'Dynamics of photoelectrons and structural changes of tungsten trioxide by femtosecond transient XAFS', Y. Uemura, D. Kido, Y. Wakisaka, H. Uehara, T. Ohba, Y. Niwa, S. Nozawa, T. Sato, K. Ichiyanagi, R. Fukaya, S.-i. Adachi, T. Katayama, T. Togashi, S. Owada, K. Ogawa, M. Yabashi, K. Hatada, T. Takakusagi, T. Yokoyama, B. Ohtani, K. Asakura, Angew. Chem. Int. Ed., 2016, **55**, 1364–1367.

42 'Multiphoton L-shell ionization of H_2S using intense x-ray pulses from a free electron laser', B. F. Murphy, L. Fang, M.-H. Chen, J. D. Bozek, E. Kukk, E. P. Kanter, M. Messerschmidt, T. Osipov, N. Berrah, Phys. Rev. A, 2012, **86**, 053423.

43 'X-ray two-photon absorption competing against single and sequential multiphoton processes', K. Tamasaku, E. Shigemasa, Y. Inubushi, T. Katayama, K. Sawada, H. Yumoto, H. Ohashi, H. Mimura, M. Yabashi, K. Yamauchi, T. Ishikawa, Nat. Photonics, 2014, **8**, 313–316.

44 'Photon-in photon-out hard X-ray spectroscopy at the Linac Coherent Light Source', R. Alonso-Mori, D. Sokaras, D. Zhu, T. Kroll, M. Chollet, Y. Feng, J. M. Glownia, J. Kern, H. T. Lemke, D. Nordlund, A. Robert, M. Sikorski, S. Song, T.-C. Weng, U. Bergmann, J. Synchrotron Radiat., 2015, **22**, 612–620.

45 'Ultrafast excited state relaxation of a metalloporphyrin revealed by femtosecond x-ray absorption spectroscopy', M. L. Shelby, P. J. Lestrange, N. E. Jackson, K. Haldrup, M. W. Mara, A. B. Strickrath, D. Zhu, H. T. Lemke, M. Chollet, B. M. Hoffman, X. Li, L. X. Chen, J. Am. Chem. Soc., 2016, **138**, 8752–8764.

46 'The Mn_4Ca photosynthetic water-oxidation catalyst studied by simultaneous x-ray spectroscopy and crystallography using an x-ray free-electron laser', R. Tran, J. Kern, J. Hattne, S. Koroidov, J. Hellmich, R. Alonso-Mori, N. K. Sauter, U. Bergmann, J. Messinger, A. Zouni, J. Yano, V. K. Yachandra, Phil. Trans. R. Soc. B, 2014, **369**, 20130324.

47 'A multi-crystal wavelength-dispersive x-ray spectrometer', R. Alonso-Mori, J. Kern, D. Sokaras, T.-C. Weng, D. Nordlund, R. Tran, P. Montanez, J. Delor, V. K. Yachandra, J. Yano, U. Bergmann, Rev. Sci. Instrum., 2012, **83**, 073114.

48 'A von Hamos x-ray spectrometer based on a segmented-type diffraction crystal for single-shot x-ray emission spectroscopy and time-resolved resonant inelastic scattering studies', J. Szlatchetko, M. Nachtegaal, E. de Boni, M. Willimann, O. Safonova, J. Sa, G. Smolentsev, M. Szlatchetko, J. A. van Bokhoven, J.-Cl. Dousse, J. Hoszowska, Y. Kayser, P. Jagodzinski, A. Bergamaschi, B. Schmitt, C. David, A. Lücke, Rev. Sci. Instrum., 2012, **83**, 103105.

5

Data Analysis and Simulation Methods

Conversion of XAFS data to the most probable physical or chemical result is achieved by a variety of methods. There are three main approaches, depending upon the problem.

1) The first approach deals with identifying known phases for which good reference spectra are available. The issue in this case is quantitative and/or spatial analysis.
2) The second application of XAFS is to probe the physico-chemical properties of the material, including electronic and magnetic properties. This too may include known materials, with spatial analysis of the properties.
3) The third approach lies in structure elucidation of an unknown material, often in which long-range order is absent. So there may be no crystallographic information about the sample, and thus the aim is to derive the maximum structural detail that has a good probability of being an accurate description. This approach is illustrated in Figure 5.1; the example information is taken from a study of a homogeneous catalyst;[1] the species was an intermediate in the activation of a paramagnetic precursor in solution. On the left of the figure are typical XAFS inputs: the XANES pattern, the EXAFS scattering pattern in k space and its Fourier transform indicating the components in R space. The EXAFS refinements provide a set of possible structural units. These may be investigated theoretically, providing a three-dimensional option and this local structure used as an input to calculations of the XANES spectrum. After a series of cycles, guided by the information on the sample from other techniques, "success" is a structural model supported by all the pieces in the puzzle and that is energetically accessible under the experimental conditions.

This chapter provides guidance in utilizing XAFS data for structural, analytical, and physico-chemical information.

There are several data reduction and analysis procedures available worldwide. Links to these have been collected until a few years ago on *xafs.org*, but this remains a useful entry point. The International X-Ray Absorption Society,

X-Ray Absorption Spectroscopy for the Chemical and Materials Sciences, First Edition. John Evans.
© 2018 John Wiley & Sons Ltd. Published 2018 by John Wiley & Sons Ltd.

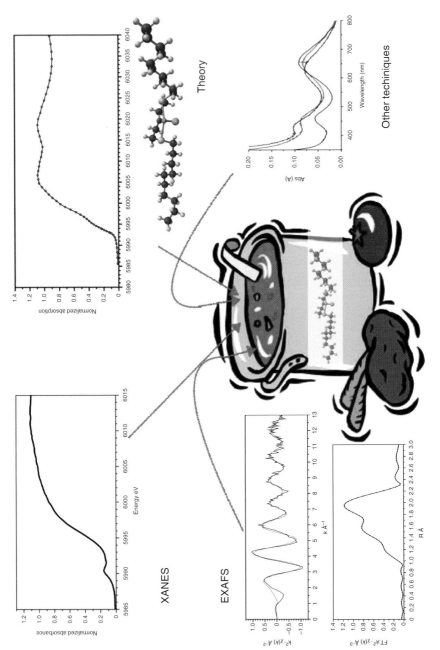

Figure 5.1 Input into arriving at chemical structures from XAFS measurements.

which organizes the series of international XAFS conferences, has its own portal: www.ixasportal.net/xas, which includes links to tutorials, books, and codes. Probably the most-used package is the *Demeter* suite (https://bruceravel.github.io/demeter), which is available in Linux, Windows, and Mac versions. The suite includes the programs ATHENA, ARTEMIS, and HEPHAESTUS.[2] ATHENA provides data processing routines and HEPHAESTUS provides x-ray data, useful for planning experiments as well as for the analysis of the results. ARTEMIS is the part of the *Demeter* package dealing with structural analysis of XAFS spectra, providing routes to FEFF, Larch, and IFEFFIT.

5.1 Background Subtraction

5.1.1 Experimental Considerations

A rather standard example of achieving a good background subtraction is illustrated in Figure 5.2, which revisits an example from Chapter 2. This was a transmission XAFS measurement on a homogeneous solution (Section 4.3.1). Detector offsets can then be acquired readily with beam-off measurements, and the x-ray flux on the sample is tracked point-by-point through the energy spectrum by the signal from the I_0 detection system, here using an ion chamber. The energy calibration can be checked with a foil between the second and third ion chambers. Figure 5.2a emphasizes the variation in the sample absorption across the spectral range. The raw spectrum is also of relative rather than absolute absorbance since the ratio of I_0 and I_t will depend upon the gas composition in each ion chamber, and the amplifier settings. Spectra are therefore normalized with an edge-step of unity (Figure 5.2b). The edge step is scaled from the difference between the pre-edge and post-edge background at an energy point above the edge when the XAFS oscillations are low or cross the post-edge background. Neither of these background curves is directly measured experimentally. However, with sufficient length of scan they can be well-defined (in this example the pre-edge region was recorded for 300 eV). That time should not be begrudged—it can influence the amplitude profile of the EXAFS features as well as define the edge-step. This is readily illustrated by Figure 5.2a; any error in the slope of the backgrounds will affect the relative magnitude of the oscillations and the edge jump.

Other measurement modes present less ideal conditions for background subtraction. In the energy dispersive mode (Section 4.2.2), I_0 is not generally measured simultaneously. In principle, a detector such as an ion chamber or photodiode could be place in front of the sample. That could only scale the whole of the band spread and not the individual energies; it would also introduce an optical element that may result in x-ray scatter and thus anomalous

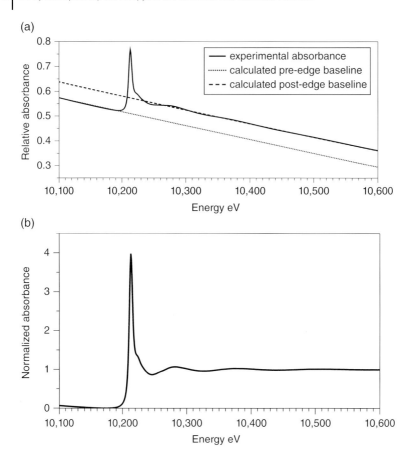

Figure 5.2 a) Experimental x-ray absorption spectrum of the L_3 edge of $(NBu_4)_2[WO_4]$ in MeCN solution showing pre- and post-edge backgrounds. b) normalized x-ray absorption (*Source*: B18, Diamond, data from Richard Ilsley).

features in the dispersed spectrum. In practice a separate I_0 reference sample is required and the sample mounting moved between the positions whereby the sample or the I_0 reference are at the beam focus. Thus the I_0 will be displaced in time by seconds or minutes. This is demanding on beam stability; the introduction of top-up modes of storage ring operation maintains an almost constant heat load on the optical components that does improve stability of the x-ray beam on the sample. The I_0 reference sample should be as closely matched to the properties of the sample without the absorption edge.[3]

The treatment of backgrounds in imaging experiments varies with the method. Full-beam imaging requires acquisition of images with the beam-off as the instrumental background, with the edge jump assessed by pixel-by-pixel differences between the images taken with the x-ray energy below and above the absorption edges. The sets of images must be extended with polarization studies to include I_0 sets at each polarization.[4] For raster mapping with Kirkpatrick-Baez focusing elements an I_0 detector can be located in front of the sample, and the situation is similar to that for a scanning spectrometer.[5] This is not feasible with a STXM given the short space between the zone plate, order-sorting aperture, and the sample. The extraction of XAFS spectra from the stack of images obtained over the energy scan in a pixel-by-pixel fashion.[6,7]

5.1.2   Background Subtraction Procedures

The data treatment as far as Figure 5.2 provides the normalized energy-calibrated spectrum, which is the first stage of the background subtraction procedures and can be the basis of compositional analysis and XANES simulations. For structural analysis, the absorption edge position, E_0, must be estimated and the EXAFS oscillations extracted. Generally the position of E_0 is taken from the maximum rising slope on the absorption edge, and thus the first derivative of the absorption spectrum with respect to energy is also calculated (Figure 5.3). This is taken as the energy zero point from which the photoelectron wave vector, k, is measured.

The magnitudes of the EXAFS features in k space are derived from the curves shown in Figure 5.4. The shape of the absorption curve is build up from the

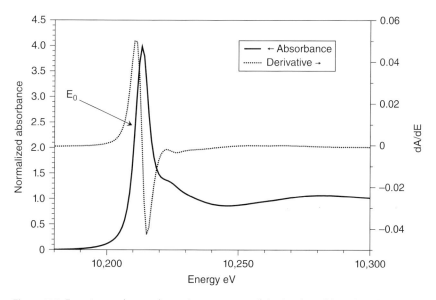

Figure 5.3 Experimental x-ray absorption spectrum of the L_3 edge of $(NBu_4)_2[WO_4]$ in MeCN and the first derivative with energy (*Source:* B18, Diamond, data from Richard Ilsley).

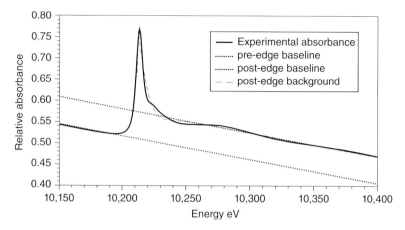

Figure 5.4 Experimental x-ray absorption spectrum of the L_3 edge of (NBu$_4$)$_2$[WO$_4$] in MeCN showing pre- and post-edge baselines and the post-edge background (*Source:* B18, Diamond, data from Richard Ilsley).

baselines, generally by a spline technique, and the EXAFS features are the difference between the experimental and post-edge background relative to the difference between the post- and pre-edge backgrounds. This is not always an easy process, and there are often stages of iteration between background subtraction and analyses. Particular problems emerge in modeling the background at low k, close to the sharp gradient changes around the edge. Also if there are some two-electron processes in the post-edge region this can afford a slope change in the post-edge absorbance. This can result in a low frequency oscillation that manifests itself in a component of implausibly low interatomic distance.[8]

The resulting EXAFS, and the Fourier transform components of the spectrum in Figure 5.4 are shown in Figure 2.13. The same process for the data reduction of (NBu$_4$)$_2$[W$_6$O$_{19}$] is shown in Figure 2.15 in terms of the k^3 weighted EXAFS and the magnitude of the Fourier transform. That k weighting is shown to amplify the back-scattering from high k values and thus emphasize the tungsten atoms. The Fourier transform approach to the analysis of EXAFS data substantially opened up the techniques to applications (equation 5.1).[9] The Fourier transform can be seen to be applied between the set range of k values on a k-weighted EXAFS as a function of the internuclear distance from the absorbing atom, R. The imaginary exponential term can be reformulated as the sum of real (cosine) and imaginary (sine) parts (equation 5.2). Thus Fourier transforms can be plotted as the sum of these (magnitude) and/or as the two components. The sine function of the EXAFS equation (equation 2.6) also includes the phase-shift associated with the photoelectron wave passing from the absorbing atom to the backscattering atom and then returning. Thus, unless a correction

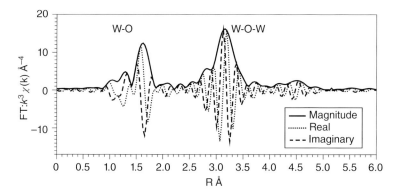

Figure 5.5 Fourier transform components of the $k^3\chi(k)$ of the W L_3 edge of $(NBu_4)_2[W_6O_{19}]$ (*Source:* B18, Diamond, data from Richard Ilsley).

term is added, the R in the Fourier transform will be in error, generally a reduction of ~0.25 Å (Figure 5.5). Nevertheless the package of absorbance, EXAFS and Fourier transform provide a rich basis for site speciation, either by comparison with models, (Section 5.2) or by structural analysis (Section 5.3).

$$\phi_n(r) = \frac{1}{\sqrt{2\pi}} \int_{k_{min}}^{k_{max}} k^n \chi(k) e^{2ikR} dk \qquad (5.1)$$

$$e^{2ikr} = cos(2kR) + isin(2kR) \qquad (5.2)$$

5.2 Compositional Analysis

5.2.1 Single Energy Comparisons

As XAFS spectra show variations in absorption that are chemically dependent, then in principle, there can be regions of the spectrum that are particularly sensitive and can be used to follow chemical change. A relatively rare example of achieving this by EXAFS, using the Pt L_3 edge, was the study of the transfer of a hydride in a heterometallic cluster onto an ethynyl ligand in $[Ru_3H(CO)_9(CCBu^t)$ $\{Pt(dppe)\}]$ (dppe = $PPh_2CH_2CH_2PPh_2$) to form a vinylidene in $[Ru_3(CO)_9\{CC(H)$ $Bu^t\}-\{(Pt(dppe)\}]$.[10] The platinum coordination changes from having one Pt-Ru bond to two, and thus there are strong EXAFS differences in the 9–11 $Å^{-1}$ region that provide a good concentration marker. It is more common that the XANES regions, with sharper features, can be used as a compositional measure. This proved useful in assessing the degree of oxidation of the heterogeneous catalyst Rh/Al_2O_3 and deriving kinetic data[11,12] (Figure 5.6). However, when the situation is more complex due to the presence of another

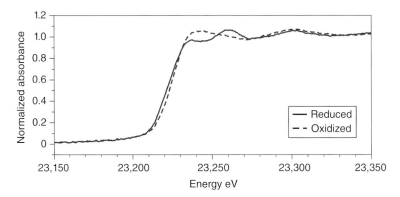

Figure 5.6 *In situ* energy-dispersive Rh *K* edge XAFS of Rh/γ-Al$_2$O$_3$ under H$_2$/He and O$_2$/He (*Source:* Data from Anna Kroner, recorded on ID24 of the ESRF).

species, as observed in *in situ* studies on CO oxidation, this type of analysis can become harder to perform. Under some conditions a third species, [Rh(CO)$_2$/Al$_2$O$_3$] is evident. Its XANES spectrum is similar to that of metallic Rh/Al$_2$O$_3$ and thus it is difficult to estimate the concentration of these components on this basis alone.[12]

5.2.2 Least Squares Analysis

An extension of this approach is to carry out a least squares analysis of the normalized spectrum of a sample with those of potential components of the sample. Figure 5.7 illustrates this approach during a study the activation of the homogeneous catalyst [CrCl$_3$(PPh$_2$N^{i}PrPPh$_2$)(THF)] with trimethylaluminium.[13] The spectrum obtained from quenching the reaction after 5 minutes differed significantly from those derived from a sample quenched after 1 minute, and acquired more conventionally at ambient temperature (Figure 5.7a). By carrying out a least squares fit of the mixture of the "1 minute" and "steady state" spectra, the proportions of these two species in the "5 minute" spectrum were estimated to be 0.49:0.51 (Figure 5.7b).

A concern with this approach is how accurate the model compound spectra reflect the components within the sample. In the example above, the spectra of the molecular components were measured in solution, or a frozen solution (as a glass rather than an ice) and were very closely related to the "unknown" sample, also studied as a frozen solution. Model spectra are often collected on polycrystalline samples or metal foils, which normally have high long-range order. In samples of interest, such as a heterogeneous metal catalyst, these components may be present as distributions of small particles and/or a mosaic of nanocrystallites. This will reduce the population of non-bonded, remote neighbors, and also increase the static disorder of shells. In addition, there is

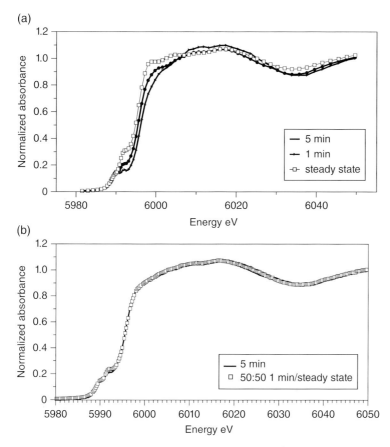

Figure 5.7 Cr K edge XAFS of the reaction of [CrCl$_3$(PPh$_2$N^{i}PrPPh$_2$)(THF)] with AlMe$_3$ in toluene. a) Comparison of spectra obtained after 1 min, 5 min, and the steady state. b) Comparison of the spectrum after 5 minutes reaction time with a 50:50 mixture of the spectra after 1 minute and the steady state (*Source:* B18, Diamond, data from Stuart Bartlett).

likely to be lower rigidity in the structure, thus increasing the Debye-Waller factors and dampening the EXAFS from these shells; that effect is likely to be much larger for the non-bonded shells in which the interatomic distances are influenced by (lower frequency) bending modes. This dampening due to increased thermal components of the Debye-Waller factors will be enhanced at increased temperatures. As a result, there is a balance to be found to optimize the accuracy of the least square fitting procedure. Increasing the data length enhances the discrimination feasible by this approach, but at higher k values, the effects of variations in static and thermal disorder can become problematic.

A series of spectra taken on related samples, for example, of different regions of a textured material or of a time sequence of chemical reaction, will contain within it the signatures of the components materials and their proportions. This can be the subject of a least squares combination fit of possible components. As illustrated by a kinetic study of the reduction of $[IrCl_6]^{2-}$ by $[Co(CN)_5]^{3-}$ using the Ir L_3 edge, success depends upon very careful and consistent normalization.[14] This showed that the first spectrum acquired in the time series was not that of the Ir(IV) precursor, which had been transformed into an inner-sphere bridged Ir-Cl-Co bridged complex within 200 ms.[15]

Another example of applying this approach is provided by an *in situ* activation of a heterogeneous catalyst for CO oxidation, which was formed by adding Rh to a mesoporous silica incorporating 5 wt % of Cr.[16] Comparison of the Rh XANES of the sample under He shows that the Rh is predominantly oxidized (Figure 5.8a). A least square analysis can be carried out on the entire series of spectra (using ATHENA) within the heating program under a reducing atmosphere (H_2/He). The XANES region was adopted for the analysis since the sample temperature was being varied over a 270° range and so the Debye-Waller factors were likely to increase through the heating cycle and thus dampen the EXAFS features of the constituents. The analysis is again based upon the standard spectra of Rh and Rh_2O_3, and provides the proportions, errors, goodness of fit figures of merit, and the variation in E_0 required for best alignment. The results in this case (Figure 5.8b) show an onset of further reduction of the rhodium sites at 390 K reaching a constant level of about 90% above 500 K.

In a significant number of studies, this approach will not provide a complete and accurate description of the sample. No standard spectrum might exist, particularly for a reaction transient. In that situation, a standard spectrum of one or more components could be could be subtracted from experimental spectrum, providing the spectrum of an unknown in the residuals. In practice the residuals will include a concentrate of the imperfections in the data.

5.2.3 Principal Component Analysis

An alternative method is by principal component analysis (PCA) of an array of spectra. The data in formed of normalized spectra set into a matrix **A** of P columns and N rows where P is the number of data points in each spectrum and N the number of spectra. This is transformed into three component matrices: the components **C**, $(P \times N)$, a diagonal matrix **E** $(N \times N)$ of the eigenvalues and the transpose of a square matrix of the weightings **W** $(N \times N)$ (equation 5.3).

$$[A] = [C].[E].\left[W^T \right] \tag{5.3}$$

The number of components is therefore identical to the number of input spectra, the eigenvalues represent the importance of each component in the entire

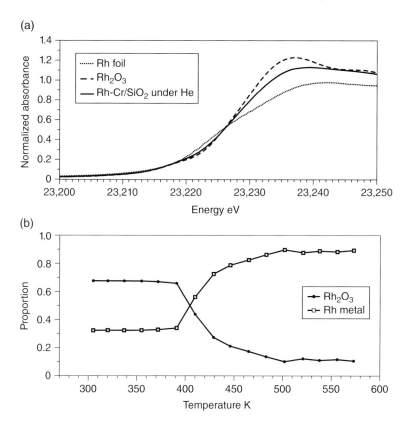

Figure 5.8 Activation of 0.5%Rh–5%Cr on mesoporous SiO₂ under H₂/He. a) Rh *K* edge XANES of the sample at 305 *K* under He with that of Rh standards. b) Proportions of Rh(0) and Rh(III) derived from least squares analyses during heating under H₂/He (*Source:* B18, Diamond, data from Khaled Mohammed).

series of data and the weightings provide the scaling of an individual spectrum to each component. The principal spectral features are emphasized in the larger eigenvalue components with the noise accumulating in the lesser eigenvalue components. Thus it may be that a few component spectra, either of model compounds or simulations from structural analysis, may be needed to provide a good fit of all the spectra in a series, but varying in the proportions of the components.

Taking 21 spectra from the experiment shown in Figure 5.8, then 21 components will be derived from this approach. As shown in Figure 5.9a, the major spectral changes are contained in the first two eigenvalues; the third component in this case can be seen to be minor. Unsurprisingly then, the spectrum of, for example, the sample under helium before the heating program can be reconstructed by two components with only a small residual. What is more structurally helpful is to compare the resultant components with the spectra of

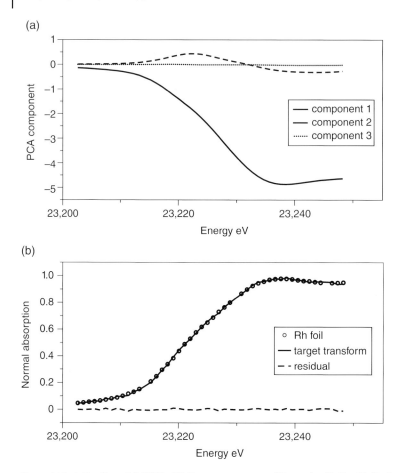

Figure 5.9 Activation of 0.5%Rh–5%Cr on mesoporous SiO_2 under H_2/He, Rh K edge spectra. a) the first three eigenvalues of the principal component analysis of the series of 21 spectra. b) comparison of the target transform with the spectrum of Rh foil (*Source:* B18, Diamond, data from Khaled Mohammed).

chemical targets, the matrix **T**, with its spectrum transformed using the eigenvectors from the experiment (**T***) (equation 5.4). Figure 5.9b shows the comparison of the transformed spectrum with that of Rh foil, a plausible model component of the series. There is a close match indicating that metallic rhodium is very likely to be a constituent in this series of spectra. A similar comparison using Rh_2O_3 also indicates as a potential constituent. Thus further analysis could proceed on the basis of these two constituents.

$$\left[T^* \right] = \left[C \right] . \left[C^T \right] . \left[T \right] \tag{5.4}$$

5.2.4 Mapping Procedures

The most obvious means of mapping heterogeneous samples beyond x-ray absorption contrast is through elemental analysis by x-ray fluorescence, akin to EDX methods in electron microscopy. Procedures are established that allow the MCA output from energy dispersive multi-element fluorescence detectors to be the source data for such mapping in the program *PyMCA*.[17] Probing in more depth to achieve chemical and structural discrimination is more demanding in energy resolution and thus in terms of sensitivity. The PCA methods can be used to group regions of materials into common constituents and thus allow summing of spectra to improve signal to noise ratios. An example is the study of the calcium-containing granules excreted by earthworms by *K* edge XANES[18] at the Microfocus Spectroscopy beamline I18 at Diamond (Figure 5.10a). In this case the μXANES analysis was aided by carrying out μXRD maps of the sample. An independent map of polymorphs of calcium carbonate could be derived from the integrated intensities of a set of diffraction lines using the *PyMCA* routine. The μXANES imaging was derived from an array of x-ray fluorescence maps by rastering the sample across the x-ray beam and collected using different excitation energies across the Ca *K* edge XANES together with the I_0 values of each measurement. Using the program *Mantis*,[19] individual normalized XANES spectra of each pixel (5 × 20 μm) could be extracted. Regions of the sample with similar spectra could then be carefully clustered together to allow for identification of the component spectra (Figure 5.10b). By comparison with reference samples, these three components can be identified as vaterite, calcite, and a mixture of the two (calcite:vaterite = 2:1) and the clusters mapped (Figure 5.10c); no aragonite or amorphous calcium carbonate could be identified in the region studied. The μXRD maps confirm the phases in these regions, but individual patterns were acquired with ~6 times the volume than for μXANES. The analysis has to be performed with careful iterations like this if it is to be vested with a high degree of confidence.

A major challenge of imaging, and also for *in situ* time-resolved studies, is to correlate and analyze the data blocks from multiple methods of measurement. For example, once our time *operando* catalysis studies started to work efficiently with a 10 Hz repetition rate, analyzing the 250,000 XAFS, IR, and mass spectrometry files required block processing. Recently, this approach has been taken to a much higher level with *in situ* spatial and time-resolved studies of single-catalyst particles; this has required a development of an automatic multimodal approach to data arrays. These are based upon Nexus format files that include the metadata essential for correlating space and time in all the data streams.[20]

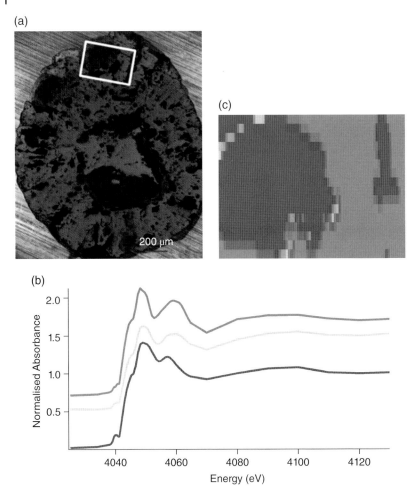

Figure 5.10 A calcium carbonate granule excreted from the earthworm *Lumbricus terestris*. a) optical microscope image showing region of μXANES and μXRD study b) the component spectra identify from the μXANES mapping: bottom component 1, vaterite; middle component 2, calcite; top component 3, mixture c) phase map: dark gray vaterite, light gray calcite (*Source:* Courtesy of Fred Mosselmans).

5.3   Structural Analysis

5.3.1   EXAFS Analysis

The EXAFS phenomenon is introduced in Section 2.1.3, and the standard equation quantifying the event for a randomly oriented sample is reproduced here for convenience (equation 5.5). This is a practical modification of that in the initial equation reported by Stern, Sayers, and Lytle.[9] Shortly afterward, this model was extended by Lee and Pendry[21] to describe the photoelectron wave as a series of harmonics, rather than maintaining the approximation of a plane wave; this can remove the approximations of small atoms and high photoelectron energies. As a result the standard approach for the last 30 years has been based on code of this nature,[22] which included the effects of multiple scattering.[23]

$$\chi(\mathrm{k}) = \sum_j S_o^2 N_j \frac{|f_j(k)|}{kR_j^2} \sin(2kR_j + 2\delta_c + \phi)e^{-2R_j/\lambda(k)}e^{-2\sigma_j^2 k^2} \tag{5.5}$$

Developments in methodology have taken place in different centers, including the codes *GNXAS*[24,25] and *FEFF*;[26–29] these developments have included the refinement procedures, the precision of calculated parameters, and reduction of the number of experimental parameters.

The model is of the excited core electron leaving the absorbing atom as a spherical wave and undergoing a phase change (δ_c) as it leaves the atom to propagate into the neighborhood. The neighborhood is modeled with a *muffin-tin* potential. The illustration of a muffin or baking tin in Figure 5.11a fails to represent this in that the muffin-tin radius is generally set so that the circles of the muffins (atoms) touch. The muffin-tin energy of the flat potential surface is the region where the wave propagates ($2kR_j$) until it is scattered back from one or more neighboring atoms, undergoing a second phase shift (ϕ). The back-scattering returns the wave to the absorbing atom where it again undergoes a phase shift (δ_c).

(a)

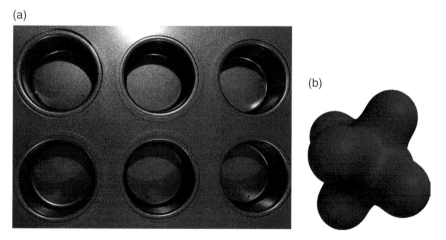

(b)

Figure 5.11 Shapes of potentials a) a muffin-tin and b) the ionization potential map calculated for [Cr(CO)$_6$] by density functional theory (Spartan'16).

A weakness in the *muffin-tin* model is the sharp cusp in the potential at the edge of the atoms. This differs markedly from the softer shapes computed from computational chemistry approaches. An example of this is a map of the ionization potential of $[Cr(CO)_6]$ (Figure 5.11b). As a result the muffin-tin approach is prone to generating low R features (~50% of the interatomic distance), and these require a critical assessment.[28]

5.3.1.1 Distance Measurement

As far as structure determination is concerned, the accuracy of the interatomic distance, R_j, is dependent upon the felicity of the calculated phase shifts. These in turn relate back to the parameters of the muffin-tin potential. Generally, for EXAFS analysis, the calculated phase shifts behave well. The other parameter that correlates with distance is E_0, which sets the start of the k region, and is close to the muffin-tin energy. In practice there is a variation in fitted values of a few eV due to both the difficulty in identifying the true E_0 from experimental spectrum and compensating for discrepancies between theory and experiment. Taking a range of different EXAFS analyses for data of "practical" quality, we estimated the precision for a single-point determination to be ~1.5%.[30] This is not dissimilar to the variation in the expectation value of a metal-ligand distance as determined by x-ray crystallography.[31] The EXAFS precision can be improved with care and with long spectral ranges; the latter reduce the correlation with E_o and enhance the accuracy of modeling the sinusoidal pattern. The resolution that can be achieved between shells of different distances is related to the k range that is analyzed (equation 5.6). An EXAFS analysis might typically begin at k_{min} of 3 Å^{-1} (34 eV beyond E_0), and a scan to 850 eV would be required attain k_{max} of 15 Å^{-1}; this gives an optimum resolution of 0.13 Å. A scan range of 2000 eV, unusually long, would afford Δk of 20 Å^{-1} and an optimum resolution of 0.08 Å.

$$\Delta R \geq \pi \big/ 2\Delta k \tag{5.6}$$

In practical terms, resolution of <0.2 Å requires lengthy data of good signal/noise. The reason for that is shown in Figure 5.12. The right-hand curves were calculated using a difference in bond length of 0.1 Å. The shorter distance gives rise to the scattering pattern with a longer oscillation in k space, as expected. Toward the end of this scan range the waves are essentially out of phase and would largely cancel each other. Thus the effect would be that this data would be fitted as a single shell with a large Debye-Waller factor. In the right hand comparison in Figure 5.2, the longer Ni-O distance (2.2 Å) is now oscillating at a higher rate in k and so the out-of-phase region is most pronounced in the 6–9 Å^{-1} region. Above 12 Å^{-1} the waves are close to being in phase and thus reinforcing each other. The effect now is that there will be an evident beat in the XAFS amplitude pointing to the need for modeling with two shells.

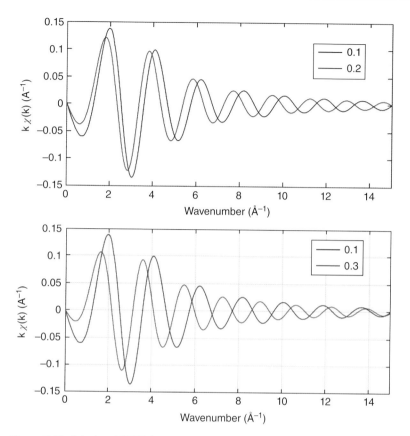

Figure 5.12 Calculated (FEFF) *k*-weighted EXAFS for two Ni-O shells. Top 2.0 and 2.1 Å; bottom 2.0 and 2.2 Å.

5.3.1.2 Angle Estimation

Equation 5.5 intrinsically contains no angular information. There are, however, two ways in which EXAFS analysis can provide estimates of bond angles.

1) Multi-edge analysis

A standard EXAFS analysis can be carried out from more than one standpoint, and thus provide angle by triangulation. An example of this is the square planar nickel complex [NiBr$_2${PPh$_2$(C$_2$H$_4$)PPh$_2$}], for which Ni and Br have accessible K absorption edges.[30] In addition to observing the metal ligand distances from both metal and ligand edges, the Br.Br and Br.P non-bonded distances could be observed and hence the Br-Ni-Br and Br-Ni-P bond angles (Figure 5.13).

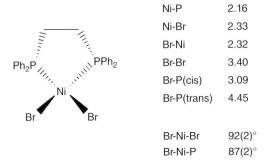

Ni-P	2.16
Ni-Br	2.33
Br-Ni	2.32
Br-Br	3.40
Br-P(cis)	3.09
Br-P(trans)	4.45
Br-Ni-Br	92(2)°
Br-Ni-P	87(2)°

Figure 5.13 Estimation of bond angles in [NiBr$_2$\{PPh$_2$(C$_2$H$_4$)PPh$_2$\}] from EXAFS analysis using the Ni and Br K edges.

2) Multiple scattering

The image of the scattering of a photoelectron wave being like a wave bouncing off a coot in a lake (Figure 1.3) only gets so far. An alternative view is like a pin-ball machine in which the electron bounces off many posts before returning to its starting point. An example of this multiple scattering is presented by a metal carbonyl (Figure 5.14). The single scattering pathway for the M...O unit involves direct scattering from the oxygen. However, there are also options for double scattering, in two different sequences, and triple scattering. Unlike a pin-ball machine where the ball cannot pass through posts, here the effect of the intervening carbon is described as a focusing one. The multiple scattering pathways each give components of the total scattering that are out of phase with each other but amplify the magnitude of the scattering (Figure 5.15 left); the multiple scattering pathways dominate the scattering involving the oxygen atom. When the bond angle is reduced, the pathways no longer have the same path length, and the phasing is more destructive.[32] As a result the magnitude of the total scattering is reduced (Figure 5.15), with the effect on the triple

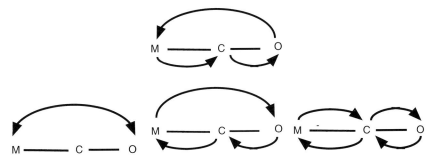

Figure 5.14 Scattering pathways for a linear M-C-O unit: single (left), double (center), and triple (right).

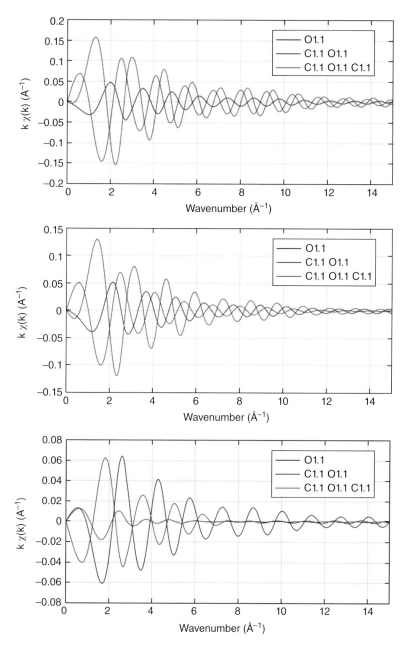

Figure 5.15 Calculated *k*-weighted Co *K* edge EXAFS of a Co-C-O unit (Co-C 1.85 Å, C-O 1.15 Å) with different bond angles: 180° (top), 150° (center), and 120° (bottom) (using FEFF).

scattering pathways being the most severe. Multiple scattering allows EXAFS to be a sensitive measure of bond angles when they are close to linear.

The importance of multiple scattering involving the absorbing atom is strongly dependent upon the type of ligands and the coordination geometry. In the complexes shown in Figure 5.16, complex **A**, $[NiBr_2(PPh_3)_2]$, is tetrahedral and there are no long-range pathways evident, even when the sample was held at 10 K.[33] In contrast, *trans*-$[NiBr_2(PEt_3)_2]$ **B** shows a distinct peak at $2R$, even at room temperature. Multiple scattering calculations indicated that the scattering pathways involving the higher atomic number ligand and those visiting the absorbing atom twice were the more intense; pathways involving *cis* ligands were negligible. The effect of bond angle is shown in Figure 5.17. For a linear Br-Ni-Br unit, the peak in the Fourier transform at $2R$ is very clear and the contributions that dominate involve scattering from both the bromine atoms. However at a Br-Ni-Br angle of 120° these contributions are virtually

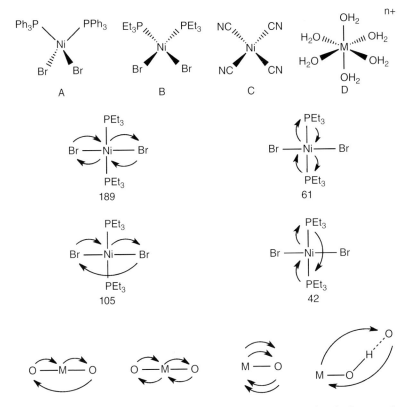

Figure 5.16 Top: four complexes illustrating the characteristics of multiple scattering. Middle: the most important multiple scattering contributions to the Ni K edge EXAFS of complex **B**. Bottom: the important scattering pathways to the EXAFS of the metal edges of octahedral aquo complexes in solution.

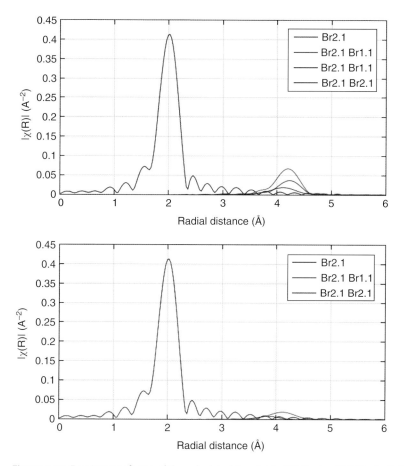

Figure 5.17 Fourier transforms of the calculated *k*-weighted Ni *K* edge EXAFS of a Br-Ni-Br unit (Ni-Br 2.298 Å) with different bond angles: 180° (top) and 120° (bottom) (using FEFF).

negligible with the "rattle" of two visits to the same bromine showing similar, low magnitude. Generally, the Debye Waller factors are larger for off-axis scattering, which also enhances the selectivity for linear units. For the complex $[Ni(CN)_4]^{2-}$, there is competition between intra- and inter-ligand multiple scattering with linear triatomic units.[34] In this case, the single and multiple scattering within the Ni-C-N units dominate. In the absence of a strongly multiple scattering unit inter-ligand higher-order pathways can be identified with low Z ligands.[35] This was achieved in the hexa-aquo complexes $[M(OH_2)_6]^{3+}$ (M = Cr and Rh) where there are only the weakly scattering hydrogen atoms bonded to the coordinating atom. The dominant multiple scattering contributions are shown in Figure 5.16, but the single scattering to the second hydration sphere is of comparable magnitude; both types of pathway must be considered to model the scattering. The electron mean free path (λ)

also affects intensity and has some *k* dependence (as the energy of the electron wave alters across the spectrum). This too is generally estimated from a fit of the spectrum of model materials.

5.3.1.3 Coordination Number Estimation

Establishing an accurate estimate of the coordination number of a shell, N_j, is complicated by the other factors in equation 5.5 that have intensity implications, and as a result the precision is generally quoted as ±10%. This is can be optimistic, especially for shells of similar distance. A direct complication is the term S_o^2, the probability of an absorbed x-ray creating XAFS. Both these terms are independent of *k* and thus are completely correlated. Empirical values are generally in the range of 0.8 ± 0.1. This is normally estimated by carrying out EXAFS analysis on a known (model) material of known structure and coordination number. Thus the N_j values are fixed and S_o^2 and $\lambda(k)$ of a similar material and the same absorption process can be refined, and then used as fixed parameters for samples of unknown coordination number. Theoretical values of these parameters are also available,[30] but it is always good practice to get to know the inter-relationships between parameters by fitting model materials close in nature to the problem of interest.

The remaining intensity function is the Debye Waller factor, σ_j, which appears in the exponential $e^{-2\sigma_j^2 k^2}$. The effect is shown in Figure 2.14b. An increased σ value will reduce the scattering to an extent that strongly increases with *k*. So in principal this can be disentangled from the coordination number, although there can be a high (negative) correlation between the two. As a result, there have been considerable efforts to model the Debye Waller factor, which is complicated in having both static disorder and structural dynamics components. These are modeled as symmetric Gaussians, as for a simple harmonic oscillator. Static distributions can differ greatly from that model. The dynamic factor is generally affected by anharmonicity. This is illustrated in Figure 5.18. Clearly, rising up the potential energy curve from the equilibrium bond length will give very unsymmetrical displacements.

This non-Gaussian distribution has been shown to have dramatic effects on the apparent coordination number, and also affect the refined interatomic distance;[36,37] for example, the temperature dependence of the nearest-neighbor distance in metallic zinc was found by EXAFS to contract by 0.09 Å from that at 20 *K*, rather than expand by 0.05 Å as indicated by diffraction methods;[36] similar effects were also observed in ionic conductors like the Group 11 halides [37].

This has been addressed by *cumulant expansion* methods.[28,38,39] The shape of the potential *V(r)* in modeled by an expansion of higher orders of the instantaneous interatomic distance, *r* (equation 5.7):

$$V(r) = \frac{1}{2}ar^2 + br^3 + cr^4 \ldots \ldots \tag{5.7}$$

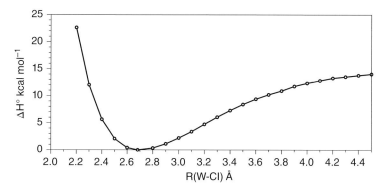

Figure 5.18 Calculated energy profile (DFT) of the W-Cl distance in the adduct [W(CO)$_5$(ClC$_6$H$_{11}$)].

In terms of the interatomic distance, R, derived from the EXAFS equation (equation 5.5), there is a shift from the equilibrium distance (R_0) due to the anharmonicity of ΔR (equation 5.8). This is the first order cumulant, C_1, and it is the odd-order (in k) cumulants that contribute to the EXAFS phase, as in equation 5.9, where ϕ_r and ϕ_s are reference and sample phases and ΔC_n, the differences between the cumulants of reference and sample.[40]

$$\Delta R = R - R_0 = C_1 \tag{5.8}$$

$$\phi_s(k) - \phi_r(k) = 2k\Delta C_1 - \frac{4}{3}k^3\Delta C_3 + \frac{4}{15}k^5\Delta C_5 + \dots \tag{5.9}$$

Thus the cumulant expansion method can add extra parameters to be refined from EXAFS analysis, and would provide a correction to the value determined for the equilibrium interatomic distance allowing for variations from non-Gaussian. Most commonly, this affect has been neglected. Probably, the effects of its neglect have been minimized by careful referencing, which includes close temperature matching between reference and sample. The relationships between the cumulants and the parameters defining the potential in equation 5.7 comprise a set of analytical equations that include temperature.[41] The effects of ignoring the cumulants have been analyzed in a few cases, including for the iodine K edge EXAFS in AgI:Ag$_2$MoO$_4$ glasses.[40,42] Using only the first-order cumulant ΔC_1, a value for ΔC_1 of -0.04 Å was determined, but with inclusion of a third-order correction, ΔC_1 was reduced to -0.02 Å; little further change was derived from inclusion of ΔC_5. So ideally, four parameters, E_0, R, C_1, and C_3 are required to properly determine the value of R; there is likely to be a systematic shortening comparable to the normal error ranges for R at ambient temperature, but the effect does increase with temperature.

The even cumulant terms affect the amplitude (A) of the EXAFS curves, with the zero order term being the coordination numbers, (N). C_2 is identical to σ^2, which is corrected with further terms in the series (equation 5.10).

$$ln\left(\frac{A_s(k)}{A_r(k)}\right) = ln\left(\frac{N_s}{N_r}\right) - 2k^2\Delta C_2 + \frac{2}{3}k^4\Delta C_4 - \frac{4}{45}k^6\Delta C_6 + \ldots \qquad (5.10)$$

The sixth order term is significant in some but not all cases [40,41], so this means that extra parameters would be required to establish the coordination number. However the number of independent parameters (N_{idp}) is limited by the range of the data in k that is used in the fit, and also the distance range that the analysis is covering, according to the Nyquist sampling criterion (equation 5.11)

$$N_{idp} \approx \frac{2.\Delta k.\Delta R}{\pi} \qquad (5.11)$$

A practical example of a Fe K-edge (7112 eV) spectrum recorded over ~1000 eV might include a 300 eV pre-edge background, and useful EXAFS data to 7750 eV (13 Å^{-1}). The first 3 Å^{-1} (to 7147 eV) may prove very difficult to fit by EXAFS programs, and thus Δk will be 10. Looking at the raw Fourier transform, there may be peaks evident from $1 - 3$ Å. In that situation, N_{idp} is ~ 12–13. If S_o^2 has not been fixed to a value from a reference compound, then, with E_o requiring to be refined, that would leave a maximum of 10 structural parameters and a distance resolution of ~0.16 Å (From equation 5.6). Using R, N and σ^2 as variables for each shell, N_{idp} would impose an absolute limit of ~3 shells. With an extra two cumulants each for distance (C_1 and C_3) and amplitude (C_4 and C_6) that would provide 7 variables per shell, limiting the fit to a single shell. Suspicions are aroused when the Nyquist N_{idp} is approached, rather than only when it is breached. It is a rare system that affords a distribution of shells evenly over the R range.

5.3.1.4 Speciation of Back-Scattering Elements

X-ray scattering factors are element specific and provide the main basis of discriminating between back-scattering elements. There are a few other strong indicators too. In some cases, the refined distance can be outside the range ever reported the two elements in question, an obvious cause for concern. Also, the change in E_0 may drift to a large value, suggesting that the wrong element has been chosen. An example of the back-scattering of two elements close to each other in the periodic table is provided by carbon and fluorine in Figure 5.19. The elements display a similar intensity pattern across the range from 2–15 Å^{-1}. The higher atomic number element, fluorine, displays markedly higher back-scattering and thus it requires two carbon atoms to roughly match the back-scattering of fluorine. The phase shifts of the two elements also differ across the k range and the EXAFS retains that differential. With good

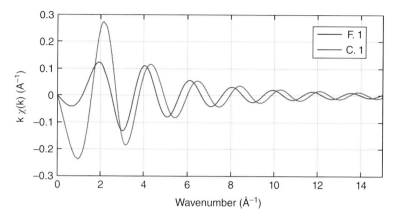

Figure 5.19 Calculated (FEFF) *k*-weighted Ni *K* edge EXAFS for Ni-F (*N* = 1) and Ni-C (*N* = 2) shells (*N* = 1) Both with bond length of 2.0 Å).

data, aided by ancillary analyses, differentiating between these two elements in the middle and end of the first short row could be feasible. It is much more difficult discriminating between elements in the same row by deeper in the periodic table.

Comparisons of back-scattering between rows are presented in Figure 5.20. It is evident that the back-scattering from higher *Z* elements is larger overall and continues to higher *k*. But there are other characteristics that are important. The amplitude of the Ni-Cl shell (Figure 5.20a) builds up more slowly than the Ni-F shell at low *k* and the shells are anti-phased for bulk of this quite normal spectral range. It can be quite demanding to acquire data good enough (in terms of signal/noise and *k* range) to reliably pick out such out-of-phase contributions that at least differ in relative amplitude across the spectrum; in a real example, the phases would differ due to the differing covalent radii of these two halogens. The amplitude envelope for Ni-Br is more complex over the first half of this *k* range and then the back-scattering clearly will be apparent beyond 15 Å^{-1}.

The back-scattering properties of nickel and palladium (Figure 5.20b) are as might be expected as they straddle bromine in the periodic table; each show dips in amplitude at low *k* and the higher *Z* element displays higher back-scattering over a longer range. The two curves move in and out of phase, indicating that it should be expected to be able to define both shells even if they were within the same sample. With two different platinum group metals as neighbors, a different characteristic emerges, not helped by the lanthanide contraction that results in the radii of *4d* and *5d* metals being very similar. For a substantial part of this *k* range the backscattering form Pd and Pt are similar in amplitude and out of phase (Figure 5.20c). Both can evidently be distinguished

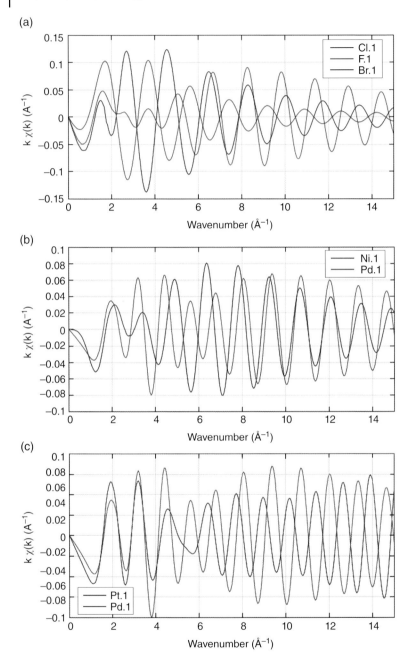

Figure 5.20 Calculated (FEFF) K weighted Ni *K* edge for a) Ni-X, all at a Ni-Cl distance (2.14 Å), b) Ni-Ni (2.50 Å) with Ni-Pd (2.63 Å), c) Ni-Ni (2.50 Å) with Ni-Pt (2.63 Å), d) Ni-C (2.00 Å) with Ni-Pt (2.63 Å).

(d)

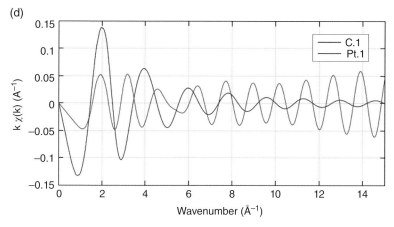

Figure 5.20 (Cont'd)

separately, but good, long k data is required to pick them out in the same sample. Another potential problem is to identify a light atom shell in the presence of a high Z backscatter, exemplified by Ni-C and Ni-Pt shells (Figure 5.20d). This shows the value of fitting with different k weightings to act as a coarse filter to emphasize elements of widely varying Z. The low k weighting offers the prospect of identifying the carbon shell through its higher amplitude at low k.

5.3.1.5 Goodness of Fit

When fitting experimental data to a structural model, the most evident rule of thumb is that the fit in k and R space should look right with a variety of k weightings. That closeness of fit is quantified as an R factor, which is the fractional difference between the experimental and theoretical EXAFS over the experimental EXAFS, or the equivalent transformed into R space. Obviously a lower R factor indicates a better fit, but this can only be quantitatively compared if the k range is unchanged in the set of fits. However, adding more shells and parameters can improve the fit, but may well not be valid statistically: a vast number of parameters should give a better fit but will violate the Nyquist criterion (equation 5.11). As a way of compensating for this, the goodness of fit is generally assessed in term of χ_v^2[43] (equation 5.12). Thus as the number of variables (N_{var}) is increased the anticipated improvement in intrinsic error estimation, χ^2, is tensioned. Clearly if $N_{var} > N_{idp}$ then χ_v^2 becomes negative, a flag for there being a formally over-determined (invalid) fit. The N_{idp} employed can often be reduced by restraining values, for example, by defining a commonality of Debye-Waller factors to a set of shells, or setting a distance-dependent relationship for them.

$$\chi_v^2 = \chi^2 \Big/ v \quad \text{where} \quad v = N_{idp} - N_{var} \tag{5.12}$$

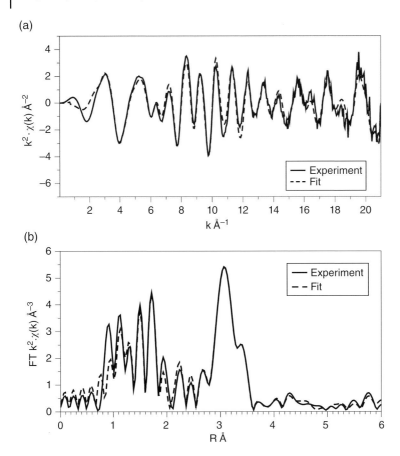

Figure 5.21 Mo K edge a) EXAFS and b) Fourier transform of $[N(PPh_3)_2]_3[Mo_{12}PO_{40}]$ (100 mM) in acetonitrile. (*Source:* B18, Diamond, data from Richard Ilsley).

The parameter χ^2 itself goes beyond the R factor and is sensitive to the data length (N_{data}) and quality (through ε, the measurement uncertainty) and also includes N_{idp}. In terms of R space fitting then χ^2 as expressed in equation 5.13 is minimized in a non-linear least squares procedure.[2,43]

$$\chi^2 = \frac{N_{idp}}{\epsilon N_{data}} \sum_{i=min}^{max} \left[Re\left\{\chi_d\left(r_i\right) - \chi_t\left(r_i\right)\right\}^2 + Im\left\{\chi_d\left(r_i\right) - \chi_t\left(r_i\right)\right\}^2 \right] \quad (5.13)$$

In addition to achieving the best statistically valid fit possible, there are also a number of additional factors to assess the plausibility of the fitting parameters:

1) S_0^2 should be in a range of 0.8–1.0
2) The change in E_0 should be only a few eV from 0

3) R values should be physically sensible
4) σ^2 values should be positive. Negative disorder is implausible and points to more sources of amplitude than are contained within the fit.
5) Standard errors associated with a parameter should be significantly lower than the value itself and not take the parameter into a danger zone.

This is to avoid inappropriate models for fitting the EXAFS data. It is nevertheless possible to arrive at satisfying conclusions. The Keggin ion, $[Mo_{12}PO_{40}]^{3-}$, presents a very favorable case. All molybdenum atoms are symmetrically equivalent in a cluster with T_d symmetry, virtually eliminating static disorder. Each Mo presents a distorted octahedron with one Mo = O unit, four Mo-O-Mo bridges and a longer Mo-O-P bridge to the central phosphate. The high energy of the Mo K edge (20 keV), and the absence of any absorption edge to high energy mean that a long data range is feasible (Recorded to 2 keV above the edge, 23 $Å^{-1}$). The rigidity of the cluster also reduces dynamic disorder, and thus shells are evident to over 6 Å. As a result $N_{idp} \approx 57$ and the resolution in R of < 0.1 Å, giving relatively sharp features in the Fourier transform (Figure 5.21).

5.3.2 XANES Simulations

5.3.2.1 *K* Edge XANES

The inclusion of higher-order multiple scattering pathways than included in EXAFS modeling was shown to be a means of calculated the XANES features.[44–46] Taking advantage of the longer electron mean free path lengths at low photoelectron energies and the sensitivity of multiple scattering pathways to local symmetry (Figure 2.8), evidence can be apparent for shells that are not observable by EXAFS.[47] This approach has been developed into simulation[48] and refinement methods,[49] and forms a basis of current treatment of near-edge spectroscopies.[29,50] An example of its application is in delineation of possible structural alternatives using XANES in the study of activation of a chromium catalyst for selective oligomerization of ethene using FEFF9 (Figure 5.22).[13] Careful calibration using the onset of the absorption of pre-edge features to give a comparison on a known structure was required to provide the energy scale. It was evidence that the Cr(II) center provided a better match with experiment that a Cr(III) alternative.

An alternative approach has stemmed from chemical modeling methodologies, with density functional theory providing the core. The stepping stones to this were calculations carried on molecular species by SCF Xα MSW codes, for example, on $Mo(CO)_6$.[51] The StoBe-deMon package[52] has provided a means of simulating XANES spectra using density functional with transition potential theory (TP), and this has been applied extensively particularly to low Z elements.[53,54] The ORCA package[55] includes time-dependent density

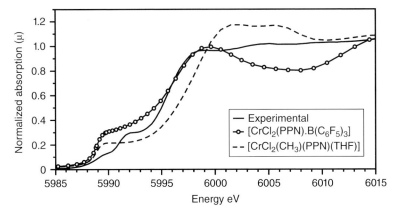

Figure 5.22 Cr K edge XANES from the reaction of [CrCl$_3$(PPN)(THF)] {PPN = (PPh$_2$)$_2$N^{i}Pr}, B(C$_6$F$_5$)$_3$, and AlMe$_3$ in toluene. Comparison of experimental spectrum calculated spectra of two possible models (*Source:* Courtesy of Moniek Tromp).

functional theory (TDDFT), which has been applied to $3d$[56] and $4d$[57] elements. The DFT-based methods have their strengths in structure prediction and in the calculation of pre-edge features. The transitions involved in these features can also delineated. This is elegantly shown in the comparison of the low spin complexes ferro- and ferri-cyanide (Figure 5.23).[56] The $3d^5$ Fe(III) center has a hole in the t_{2g} set, which gives rise to the extra pre-edge transition, formally dipole forbidden. The other features are a combination of transitions to the higher-lying e_g set of $3d$ orbitals with Fe-C σ^* character, and also charge-transfer to states primarily of CO π^* character.

The scattering-based methods have been refined in FEFF9,[29] and as a result, the differences between their results with those using DFT-based methods are reducing. An example is provided in Figure 5.24, which shows the N K edge spectra of NH$_4$NO$_3$ calculated by FEFF9 and StoBe. The chemical shift between the two nitrogen centers (4.3 eV) can now be calculated within FEFF9.[29]

5.3.2.2 L Edge XANES

In L_2 and L_3 edge spectra of the $3d$ metals, it was recognized with the first measurements that the spectral features could not be replicated solely on the basis of the splitting of these edges into two (intensity ratio 1:2) by the $2p$-$2p$ LS spin orbit coupling and the density of empty states.[58] The Hamiltonian for the interactions is given in equation 5.14.[59]

$$H = H_{av} + \mathbf{L} \cdot \mathbf{S}(p) + \mathbf{L} \cdot \mathbf{S}(d) + g(i,j) \qquad (5.14)$$

The term H_{av} provides the average energy of the spectrum. The spin-orbit vector products split this line, as does the two-electron Coulomb repulsion term (g). The Coulomb repulsion operator has a radial component, $R^K(l_1l_2;l_3l_4)$,

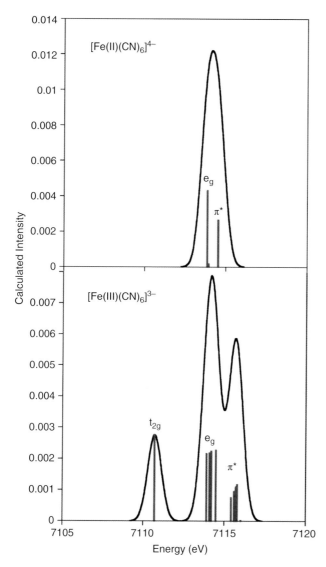

Figure 5.23 Fe *K* edge pre-edge features calculated for $[Fe(CN)_6]^{n-}$, n = 3,4 using ORCA (*Source:* George (2008).[56] Reproduced with permission of American Chemical Society).

made up of direct Coulomb $\{F^K(l_1l_2;l_1l_2)\}$ and exchange $\{G^K(l_1l_2;l_2l_1)\}$ terms, and an angular component governs selection rules that are expressed as the allowed values of the integer K. The allowed values of K for these rules:

1) For F^K, K cannot be odd, and the maximum of K is twice the minimal l value,
2) For G^K, K takes values $|l_1\text{-}l_2|$, $|l_1\text{-}l_2 + 2k| \ldots |l_1 + l_2|$

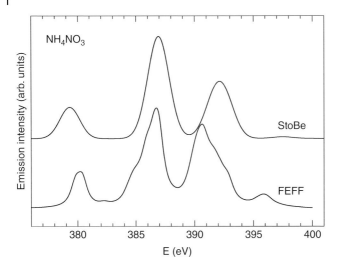

Figure 5.24 Calculated N K edge of NH_4NO_3 with FEFF9 and StoBe (*Source:* Rehr (2010) [29]. Reproduced with permission of Royal Society of Chemistry).

For a $3d^0$ ion in a $2p^5 3d^1$ excited state, these rules yield four parameters (Slater integrals): F^0, F^2, G^1, and G^3. F^0 only affects the average energy, and so is amalgamated with the average energy. Accordingly, the relative energies of the transitions of the L_2 and L_3 edges are described by one Coulomb (F^2), two exchange (G^1, G^3) and two spin-orbit coupling values $\boldsymbol{L} \cdot \boldsymbol{S}(p)$ and $\boldsymbol{L} \cdot \boldsymbol{S}(d)$.

Excitation at the $L_{2,3}$ edges ($2p^6 3d^0$, ^1S) will result in a $2p^5 3d^1$ (^1P) excited state, with the $2p$-$2p$ LS spin orbit coupling creating a pair of singlets in the Russell Saunders approach. With jj coupling, the ^1P Russell-Saunders state maps to ^1P, ^3P, and ^3D states, the triplet states being degenerate. As a result the L_3 edge is split into a weaker peak (predominantly the triplet state) in addition to the main edge (Figure 5.25).

The $2p^5 3d^1$ configuration contains 60 possible microstates, there being 6 l and s combinations that can be adopted by the $2p$ hole, and 10 for the $3d$ electron. With Russell Saunders coupling this would provide six states: ^1F, ^3F, ^1D, ^3D, ^1P, and ^3P. In an octahedral environment, the reduction in symmetry from spherical only allows a maximum degeneracy of a triplet, which for the L values of 1, 2, and 3 yield:

$$P \Rightarrow T_1; D \Rightarrow E + T_2; F \Rightarrow A_2 + T_1 + T_2$$

Transitions are only symmetry allowed if they have the same irreducible representation as the dipole vectors, namely T_1. So this treatment would predict two transitions to spin singlet states (from P and F) as the dominant features, plus possibly two very weak transitions to the triplet states.

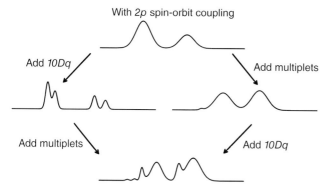

With *2p* spin-orbit coupling

Add *10Dq*

Add multiplets

Add multiplets

Add *10Dq*

Figure 5.25 Effect of the parameters contributing to the appearance of the L$_{2,3}$ absorption edges of 3d^0 ions in an octahedral field. Top: with *2p* spin-orbit coupling; center with multiplet calculations, or including an O_h crystal field; bottom, including both.

Under the *jj* coupling scheme the 60 microstates are contained within 5 irreducible representations (degeneracies) given by *J* values of 0 (singlet), 1 (triplet), 2 (quartet), 3 (triplet), and 4 (singlet). The splitting of the spherical harmonics under an octahedral field (*10Dq* or Δ_{oct}) is the same for *J* as for *L* of the same value. Hence transitions are allowed (i.e., are to T_1 states) to *J* = 1, 3, and 4. There are three degenerate states for each of *J* = 1 and 3 and one to *J* = 4, giving a total of seven transitions (Figure 5.25). It is this that is close to experimental observations; the ligand-field induced splittings in the XAFS spectrum result from, but are not equal to the energy of *10Dq*.[59] This approach was also developed for the other $3d^n$ ions, which require Coulomb term (F^4)due to the *3d* electrons.[60]

For *4d* transition metals there are some significant differences [61,62]. Firstly the larger spin-orbit coupling for the *2p* electrons means that the L_3 and L_2 edges are further apart (~100 eV). The stronger metal-ligand covalent bonding of the *4d* elements reduces the values of the *F* and *G* parameters. It also renders all complexes low spin, and thus spin states of singlet and doublet predominate. Multiplet effects are more evident in L_3 than L_2 spectra, and are manifest mostly in intensity transfer between the t_{2g} and e_g bands.

It was recognized that the intensity variations of *L* edge spectra (and also with smaller changes in *K* edge spectra) could be used as a probe for the orbital contribution to the total magnetic moment of a magnetic material, and the ratio of orbital/spin moments extracted from XMCD spectra.[63,64] These variations are angle-dependent and thus can be used to good effect to study layered materials. In a randomly oriented sample, which includes a local site with magnetic anisotropy (i.e., with non-cubic symmetry), as in a microcrystalline powder, or a solution, then a simple relationship between the spectra and the magnetic moments on an element has been established,[65,66] and its

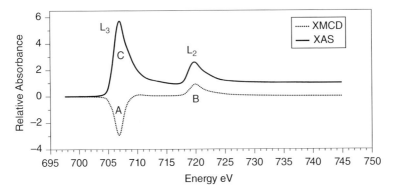

Figure 5.26 Fe $L_{2,3}$ XAS and XMCD of an iron oxide showing the peaks A, B, and C. XAS absorbance taken as the mean of positive and negative polarization values (*Source:* Courtesy of Sarnjeet Dhesi).

use in chemistry and biochemistry assessed.[67] This method is illustrated in Figure 5.26. The angular $\langle L_\zeta \rangle$ and spin $\langle S_\zeta \rangle$ angular momenta can be estimated from the intensity of XMCD signals at the L_3 (A) and L_2 (B) edges, scaled against the intensity of the L_3 edge (equation 5.15). Difficulties arise for early *3d* transition series elements when these two edges interact strongly, but the method is valuable for the later members of that series.[67] The term n_h refers to the number of holes into which the transition can occur. Without an absolute measure of intensity per absorber it is the ratio that is straightforward to extract. Values for n_h can estimated from white line areas when compared to model materials. An example of this is the study of the oxidation states of complexes of nickel, using the metal itself, whose characteristics are already established, as the reference point for intensity and n_h.[68]

$$\langle L_\zeta \rangle = \frac{2(A+B)}{3C} n_h, \quad \langle S_\zeta \rangle = \frac{A-2B}{2C} n_h \tag{5.15}$$

5.3.3 XES and RIXS Simulations

Tackling a difficult problem exhaustively can provide invaluable insights into how to develop approaches bringing together experiment and theory. One of these is the study of complexes that form models of the enzyme, *hydrogenase*, in which there are Ni-Fe bimetallic sites.[69] Protonation of [{μ-S(CH₂)₃S}{Fe(CO)₃Ni(dppe)}] (dppe = PPh₂CH₂CH₂PPh₂) at the metals will afford [{μ-S(CH₂)₃S}{HFe(CO)₃Ni(dppe)}]⁺ and an extremely challenging question is how to locate the hydride. Unsurprisingly, ORCA-based calculations failed to identify this using either the Fe or Ni *K* edge XANES. Both edges did show shifts in edge features to higher energy, consistent with a degree of oxidation as the proton is reduced to a metal hydride ligand. The $K\beta_{1,3}$ emission spectra for

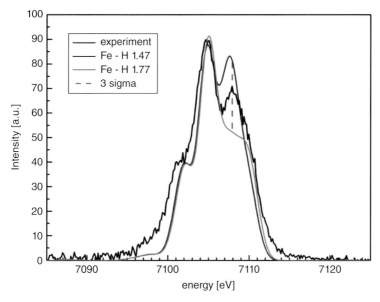

Figure 5.27 Fe *K* edge VtC emission of [{μ-S(CH₂)₃S}{HFe(CO)₃Ni(dppe)}]⁺ showing experiment, with a 3σ scaling bar, and calculations for two Fe-H distances using ORCA (*Source:* Hugenbruch (2016).[69] Reproduced with permission of Royal Society of Chemistry).

both elements displayed negligible variation on protonation. The valence to core (VtC) emissions are a few eV below the energy of the absorption edge are about 1000 times less intense, but have the advantage of involving transitions from some orbitals with M-H character. Simulations of potential structures showed a strong dependence on the M-H distance; as shown in Figure 5.27, these indicated a value of the Fe-H distance as being between 1.47 and 1.77 Å, similar to that reported by x-ray diffraction. Presently such studies are close to experimental limits.

Calculations of HERFD and RIXS patterns of the *5d* element osmium associated with the L_3 absorption have also been reported[70] using DFT-based codes and also FDMNES.[71,72] FDMNES incorporates the finite-difference method and avoids the muffin-tin approximation, which is a computationally demanding process. Recording spectra using HERFD was essential to achieve sufficient resolution to resolve a prominent post-edge transition, for example, for the octahedral complex [OsCl(bipy)₂(CO)]⁺ (bipy = 2,2' bipyridyl) (Figure 5.28a). Using the same functional (OPBE), the FDMNES approach did provide a closer match to the experimental HERFD spectrum that the DFT method employed (ADF), but the latter did provide a semi-quantitative description. All-electron DFT calculations (i.e., avoiding the use of effective core potential basis sets) do also allow the calculation of the $L\alpha_1$ RIXS spectra (Figure 5.28b).

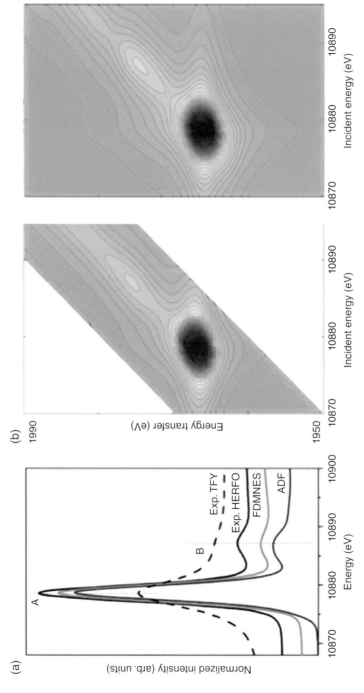

Figure 5.28 Experimental and calculated spectra of [OsCl(bipy)$_2$(CO)]$^+$ a) Os L_3 HERFD, b) Os $L\alpha_1$ RIXS, experiment (left) calculations using the ADF DFT method (right) (*Source:* Lomachenko (2013).[70] Reproduced with permission of Royal Society of Chemistry).

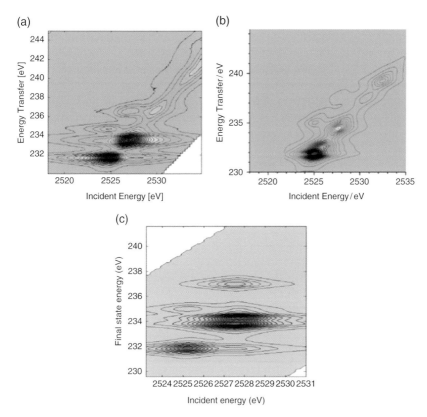

Figure 5.29 Mo $L\alpha_1$ RIXS of Na_2MoO_4. a) experimental, b) calculated with FEFF9, c) calculated with CTM4XAS multiplet code (*Source:* Thomas (2015).[73] Reproduced with permission of American Chemical Society).

The $L\alpha_1$ RIXS of the *4d* element, molybdenum in the salt Na_2MoO_4 have been calculated using FEFF9.[73] In this case, with smaller values of *2p* spin-orbital coupling as compared to the *5d* element, osmium, there is more discrepancy between calculations and experiment. This is particularly evident with underestimation of the off-diagonal peaks (Figure 5.29). These were identified as being due to multiplet splitting, and were simulated using the multiplet theory program CTM4XAS.[74]

5.4   Present To Future Opportunities

The trend described through the last few sections of this chapter provides a clear convergence of aims for the analysis of XAFS data to derive atomic and electronic structural descriptions of the samples. A selection of methods for modeling disorder is being gelled into the XAFS analysis suites, for example, in FEFF9,[29] which match the difference environments in metallic and molecular species and minimize the number of refined parameters. The development of DFT-based procedures for XAFS analysis[52,55] links much more widely used packages with intrinsic structural refinement. The same packages provide simulations of other spectroscopies (IR, Raman, uv-visible, NMR, ESR), which can provide holistic approaches to structural delineation. In that approach, XAFS provides vital checks on the bounds of theoretical fancy. Judicious application of theory can expand the structural detail beyond the scope that provided by EXAFS alone,[1] and also stimulate the acquisition of demanding but informative techniques using x-ray emission detection.[69] XAFS could be said to be entering its third age with more life skills and a greater sense of authority!

5.5   Questions

1   The Cr K edge EXAFS and Fourier transform of the reaction product observed first from the reaction of $[CrCl_3(NSS)]$ (Figure 5.1) are enlarged below.

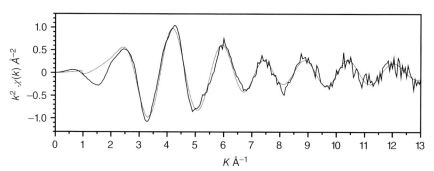

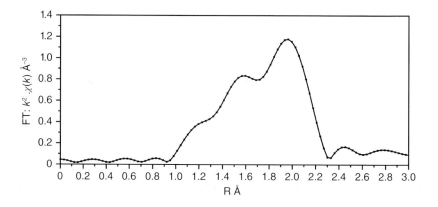

2   Identify the useful k and R ranges from these figures, and thus estimate the anticipated resolution in R, the maximum working number of independent parameters (N_{idp}), and the maximum number of shells that can be fitted. The coordination shell of chromium is thought to contain N, Cl, and S. How could you approach this analysis?

3   The dampening of the EXAFS oscillations with k by static and thermal disorder when modeled as a Gaussian function is proportional to $e^{-2\sigma_j^2 k^2}$. Plot this function up to $k = 20$ Å^{-1} for $\sigma = 0.03$, 0.06 and 0.1 Å. The experimental spectrum in Q1 could be fitted with σ values ≤ 0.03 Å. What would be the consequences for the analysis if that value had been ≥ 0.06 Å?

4   For the ion $[PtCl_6]^{2-}$, identify the single and multiple scattering pathways that could contribute to the EXAFS of the Pt L_3 edge up to a path length of 3R(Pt-Cl). Suggest which would be the most important contributors. (You can check this with the ATOMS program in the *Demeter* suite.)

5   The $L_{2,3}$ edge spectra of complexes of the early *3d* metals increase in complexity as the periodic table is progressed. Calculate the number of microstates of the ground and excited states for d^0 and d^1 ions. Account for the increase in complexity up to Mn(II). Why do the spectra simplify through the latter half of the *3d* block?

6   The Cu $K\alpha_1$ RIXS spectra of CuO (Figure 4.32b) show peaks along the diagonal in the energy transfer plot, suggesting the features are dipole-allowed. Consider the electronic configuration of the ground and excited states of the $K\alpha_1$ RIXS process, and account for the absence of multiplet features.

References

1   'Activation of $[CrCl_3\{R-SN(H)S-R\}]$ catalysts for selective trimerization of ethene: a freeze-quench Cr K-edge XAFS study', S. A. Bartlett, J. Moulin, M. Tromp, G. Reid, A. J. Dent, G. Cibin, D. S. McGuinness, J. Evans, ACS Catal., 2014, **4**, 4201–4204.
2   'ATHENA, ARTEMIS, HEPHAESTUS: data analysis for x-ray absorption spectroscopy using IFEFFIT', B. Ravel, M. Newville, J. Synchrotron Radiat., 2005, **12**, 537–541.
3   'Particle development and characterization in $Pt(acac)_2$ and $Pt(acac)_2/GeBu_4$ derived catalysts supported upon porous and mesoporous SiO_2; effect of the reductive environment, and support structure', S. G. Fiddy, M. A. Newton, T. Campbell, A. J. Dent, I. Harvey, G. Salvini, S. Turin, J. Evans, Phys. Chem. Chem. Phys., 2002, **4**, 827–834.

4 'Giant and reversible extrinsic magnetocaloric effects in $La_{0.7}Ca_{0.3}MnO_3$ films due to strain', X. Moya, L. E. Hueso, F. Maccherozzi, A. I. Tovstolytkin, D. I. Podyalovskii, C. Ducati, L. C. Phillips, M. Ghidini, O. Hovorka, A. Berger, M. E. Vickers, E. Defray, S. S. Dhesi, N. D. Mathur, Nature Mater., 2013, **12**, 52–58.

5 'The application of x-ray absorption spectroscopy in archaeological conservation: Example of an artefact from Henry VIII warship, the *Mary Rose*', A. V. Chadwick, A. Berko, E. J. Schofield, A. D. Smith, J. F. W. Mosselmans, A. M. Jones, G. Cibin, J. Non-Crystalline Solids, 2016, **451**, 49–55.

6 'Confocal soft x-ray scanning transmission microscopy: setup, alignment procedure and limitations', S. Späth, J. Raabe, R. H. Fink, J. Synchrotron Radiat., 2015, **22**, 113–118.

7 'Current status of the TwinMic beamline at Elettra: a soft x-ray transmission and emission microscopy station', A. Gianoncelli, G. Kourousias, L. Merolle, M. Altissimo, A. Bianco, J. Synchrotron Radiat., 2016, **23**, 1526–1537.

8 'Atomic background in x-ray absorption spectra of fifth-period elements: Evidence for double-electron excitation edges', A. Filipponi, A. Di Cicco, Phys. Rev. A, 1995, **52**, 1072–1078.

9 'Extended x-ray absorption technique. III. Determination of physical parameters', E. A. Stern, D. E. Sayers and F. W. Lytle, Phys. Rev. B, 1975, **11**, 4836–4846.

10 'Kinetic and structural information on metal cluster rearrangement in solution from EXAFS spectra', A. J. Dent, L. J. Farrugia, A. G. Orpen, S. E. Stratford, J. Chem. Soc., Chem. Commun., 1992, 1456–1457.

11 'Oxidation/reduction kinetics of supported Rh/Rh_2O_3 nanoparticles in plug flow conditions using dispersive EXAFS', M. A. Newton, S. G. Fiddy, G. Guilera, B. Jyoti, J. Evans, Chem. Commun., 2005, 118–120.

12 'Time-resolved in situ DRIFTS/EDE/MS studies on alumina supported Rh catalysts: effects of ceriation on the Rh catalysts in the process of CO oxidation', A. B. Kroner, M. A. Newton, M. Tromp, A. E. Russell, A. J. Dent, J. Evans, Catal. Struct. React., 2017, **3**, 13–23.

13 'Activation of $[CrCl_3\{PPh_2N(^iPr)PPh_2\}]$ for the selective oligomerization of ethene: a Cr K-edge XAFS study', S. A. Bartlett, J. Moulin, M. Tromp, G. Reid, A. J. Dent, G. Cibin, D. S. McGuiness, J. Evans, Catal. Sci. Technol., 2016, **6**, 6237–6246.

14 'Analysis of time-resolved energy-dispersive x-ray absorption spectroscopy data for the study of chemical reaction intermediate states', D. T. Bowron, S. Diaz-Moreno, Anal Chem., 2005, **77**, 6445–6452.

15 'Structural investigation of the bridged activated complex in the reaction between hexachloriridate(IV) and pentacyanocobaltate(II)' S. Diaz-Moreno, D. T. Bowron, J. Evans, Dalton Trans., 2005, 3814–3817.

16 'Preparation and characterization of some nano-structured catalytic materials for low-temperature oxidation of carbon monoxide', K. H. H. Mohammed, Ph. D. thesis, University of Southampton, 2014.

17 'A multiplatform code for the analysis of energy-dispersive X-ray fluorescence spectra', V. A. Solé, E. Papillon, M. Cotte, Ph. Walter, J. Susini, Spectrochim. Acta B, 2007, **62**, 63–68.

18 'Combining μXANES and μXRD mapping to analyse the heterogeneity in calcium carbonate granules excreted by the earthworm *Lumbricus terrestris*', L. Brinza, P. F. Schofield, M. E. Hodson, S. Weller, K. Ignatyev, K. Geraki, P. D. Quinn, J. F. W. Mosselmans, J. Synchrotron Radiat., 2014, **21**, 235–241.

19 'Cluster analysis of soft x-ray spectromicroscopy data', M. Lerotic, C. Jacobsen, T. Schäfer, S. Vogt, Ultramicroscopy, 2004, **100**, 35–57.

20 'Automatic processing of multimodal tomography datasets', A. D. Parsons, S. W. T. Price, N. Wadeson, M. Basham, A. M. Beale, A. W. Ashton, J. F. W. Mosselmans, P. D. Quinn, J. Synchrotron Radiat., 2017, **24**, 248–256.

21 'Theory of the extended x-ray absorption fine structure' P. A. Lee, J. B. Pendry, Phys. Rev. B, 1975, **11**, 2795–2811.

22 'A rapid, exact curved-wave theory for EXAFS calculations', S. J. Gurman, N. Binsted, I. Ross, J. Phys. C –Solid State Phys., 1984, **17**, 143–151.

23 'A rapid, exact curved-wave theory for EXAFS calculations. 2. The multiple scattering contributions', S. J. Gurman, N. Binsted, I. Ross, J. Phys. C –Solid State Phys., 1986, **19**, 1845–1861.

24 'X-ray-absorption spectroscopy and *n*-body distribution functions in condensed matter. I. Theory', A. Fipponi, A. Di Cicco, C. R. Natoli, Phys. Rev. B, 1995, **52**, 15122–15134.

25 'X-ray-absorption spectroscopy and *n*-body distribution functions in condensed matter. II. Data analysis and applications', A. Fipponi, A. Di Cicco, Phys. Rev. B, 1995, **52**, 15135–15149.

26 'Scattering-matrix formulation of curved-wave multiple-scattering theory: Application to x-ray-absorption fine structure', J. J. Rehr, R. C. Albers, Phys. Rev. B, 1990, **41**, 8139–8149.

27 'Theoretical x-ray absorption fine structure standards', J. J. Rehr, J. Mustre de Leon, S. I. Zabinsky, R. C. Albers, J. Am. Chem. Soc., 1991, **113**, 5135–5140.

28 'Theoretical approaches to x-ray absorption fine structure', J. J. Rehr, R. C. Albers, Rev. Mod. Phys., 2000, **72**, 621–654.

29 'Parameter-free calculations of x-ray spectra with FEFF9', J. J. Rehr, J. J. Kas, F. D. Vila, M. P. Prange, K. Jorissen, Phys. Chem. Chem. Phys., 2010, **12**, 5503–5513.

30 'Bond angle determination at metal coordination centers by EXAFS', J. M. Corker, J. Evans, H. Leach, W. Levason, J. Chem. Soc., Chem. Comm., 1989, 181–183.

31 'Tables of bond lengths determined by x-ray and neutron diffraction. Part 2. Organometallic compounds and coordination complexes of the *d*- and *f*-block metals', A. G. Orpen, L. Brammer, F. H. Allen, O. Kennard, D. G. Watson, R. Taylor, J. Chem. Soc., Dalton Trans., 1989, S1–S83.

32 'EXAFS and Near-Edge Structure in the cobalt K-edge absorption spectra of metal carbonyl complexes', N. Binsted, S. L. Cook, J. Evans, G. N. Greaves, R. J. Price, J. Am. Chem. Soc., 1987, **109**, 3669–3676.

33 'The importance of multiple scattering pathways involving the absorbing atom in the interpretation and analysis of metal K-edge EXAFS data of coordination compounds', A. van der Gaauw, O. M. Wilkin, N. A. Young, J. Chem. Soc., Dalton Trans., 1999, 2405–2406.

34 'Importance of multiple-scattering phenomena in XAS structural determinations of $[Ni(CN)_4]^{2-}$ in condensed phases', A. Muñoz-Páez, S. Díaz-Moreno, E. Sánchez Marcos, J. J. Rehr, Inorg. Chem., 2000, **39**, 3784–3790.

35 'Second hydration shell single scattering versus first hydration shell multiple scattering in $M(OH_2)_6^{3+}$ EXAFS spectra', H. Sakane, A. Muñoz-Páez, S. Díaz-Moreno, J. M. Martínez, R. R. Pappalardo, E. Sánchez Marcos, J. Am. Chem. Soc., 1998, **120**, 10397–10401.

36 'The study of disordered systems by XAFS: limitations', P. Eisenberger, G. S. Brown, Solid State. Comm., 1979, **29**, 481–484.

37 'Extended x-ray absorption fine structure investigation of mobile-ion density in superionic AgI, CuI, CuBr, and CuCl', J. B. Boyce, T. M. Hayes, and J. C. Mikkelsen, Phys. Rev. B, 1981, **23**, 2876–2896.

38 'Application of the ratio method of EXAFS analysis to disordered systems', G. Bunker, Nucl. Instrum. Meth., 1983, **207**, 437–444.

39 'On the cumulant analysis of EXAFS in crystalline solids', P. Fornasini, F. Monti, A. Sanson, J. Synchrotron Radiat., 2001, **8**, 1214–1220.

40 'On the neglecting of higher-order cumulants in EXAFS data analysis', A. Sanson, J. Synchrotron Radiat., 2009, **16**, 864–868.

41 'Fitting EXAFS data using molecular dynamics outputs and a histogram approach', S. W. T. Price, N. Zonias, C.-K- Skylaris, T. I. Hyde, B. Ravel, A. E. Russell, Phys. Rev. B, 2012, **85**, 075439.

42 'Influence of temperature on the local structure around the iodine in fast-ion-conducting $AgI:Ag_2MoO_4$ glasses, A. Sanson, F. Rocca, G. Dalba, P. Fornasini, R. Grisenti, New. J. Phys., 2007, **9**, 88.

43 'The UWXAFS analysis package: philosophy and details', E. A. Stern, M. Newville, B. Ravel, Y. Yacoby, D. Haskel, Physica B, 1995, **208–209**, 117–120.

44 'XANES: Determination of bond angles and multi-atom correlations in order and disordered systems, P. J. Durham, J. B. Pendry, C. H. Hodges, Solid State Commun. 1981, **38**, 159–162.

45 'Calculation of x-ray absorption near-edge structure, XANES', P. J. Durham, J. B. Pendry, C. H. Hodges, Computer Phys. Commun. 1982, **25**, 193–205.

46 'Multiple-scattering regime and higher-order correlations in x-ray absorption spectra of liquid solutions', M. Benfatto, C. R. Natoli, A. Bianconi, J. Garcia, A. Marcelli, M. Fanfoni, I. Davoli, Phys. Rev. B, 1986, **34**, 5774–5781.

47 'Study of the XANES modeling of molybdenum compounds', J. Evans, J. F. W. Mosselmans, J. Am Chem. Soc., 1991, **113**, 3737–3742.

48 'Near-edge x-ray absorption fine structure of Pb: A comparison and theory and experiment', M. Newville, P Līviņš, Y. Yacoby, J. J. Rehr, E. A. Stern, Phys. Rev. B, 1993, **47**, 14126–14131.

49 'The *MXAN* procedure: a new method for analyzing the XANES spectra of metalloproteins to obtain structural information', M. Benfatto, S. Della Longa, C. R. Natoli, J. Synchrotron Radiat., 2003, **10**, 51–57.

50 'New developments in FEFF: FEFF9 and JFEFF', K. Jorissen, J. J. Rehr, J. Phys: Conf. Ser., 2013, **430**, 012001.

51 '*K*-edge X-ray absorption spectra in an octahedral environment. A theoretical and experimental study of $Mo(CO)_6$', F. W. Kutzler, K. O. Hodgson, S. Doniach, Phys. Rev. A, 1982, **26**, 3020–3022.

52 'StoBe-deMon', K. Hermann and L. G. M. Pettersson, M. E. Casida, C. Daul, A. Gousot, A. Koester, E. Proynov, A. St.-Amant and D. R. Salahub. Contributing authors: V. Carravetta, H. Duarte, C. Freidrich, N. Godbout, M. Gruber, J. Guan, C. Jamorski, M. Leboef, M. Leetma, M. Nyberg, S. Patchovskii, L. Pedocchi, F. Sim, T. Triguero. A. Vela, Version 3.3 (2014); www.fhi-berlin.mpg.de/KHsoftware/StoBe.

53 'Calculations of near-edge x-ray-absorption spectra of gas-phase and chemisorbed molecules by means of density- functional and transition-potential theory', L. Triguero, L. G. M. Pettersson, H. Ågren, Phys. Rev. B, 1998, **58**, 8087–8110.

54 'Combined x-ray absorption spectroscopy and density functional theory examination of ferrocene-labeled peptides', R. G. Wilks, J. B. MacNaughton, H.-B. Kraatz, T. Regier, A. Moewes, J. Phys. Chem. B, 2006, **110**, 5955–5965.

55 'ORCA', F. Neese, F. Wennmohs. Contributions from U. Becker, D. Bykov, G. Ganyushin, A. Hansen, R. Izsák, D. G. Liakos, C. Kollmar, S. Kossmann, D. A. Pantazis, T. Petrenko, C. Reimann, C. Riplinger, M. Roemelt, B. Sandhöfe, I. Shapiro, K. Sivalingham, B. Wezisla, M. Kállay, S. Grimme, E. Valeev, G. Chan, version 3.0.3 (2016); https//orcaforum.cec.mpg.de.

56 'Prediction of iron *K*-edge absorption spectra using time-dependent density functional theory', S. DeBeer George, T. Petrenko, F. Neese, J. Phys. Chem. A, 2008, **112**, 12936–12943.

57 'High-resolution molybdenum *K*-edge x-ray absorption spectroscopy analyzed with time-dependent density functional theory', F. A. Lima, R. Bjornsson, T. Weyhermüller, P. Chandrasekaran, P. Glatzel, F. Neese, S. DeBeer, Phys. Chem. Chem. Phys., 2013, **15**, 20911–20920.

58 '*2p* absorption spectra of the *3d* elements', J. Fink, Th. Müller-Heinzerling, B. Scheerer, W. Speier, F. U. Hillebrecht, J. C. Fuggle, J. Zaanan, G. A. Sawatzky, Phys. Rev. B, 1985, **32**, 4899–4904.

59 '$L_{2,3}$ x-ray-absorption edge of d^0 compounds: K⁺, Ca²⁺, Sc³⁺, and Ti⁴⁺ in O_h (octahedral) symmetry', F. M. F. de Groot, J. C. Fuggle, B. T. Thole, G. A. Sawatzky, Phys. Rev. B, 1990, **41**, 928–937.

60 '*2p* x-ray absorption of *3d* transition-metal compounds: An atomic multiplet description including the crystal field', F. M. F. de Groot, J. C. Fuggle, B. T. Thole, G. A. Sawatzky, Phys. Rev B, 1990, **42**, 5459–5468.

61 'Differences between L_3 and L_2 x-ray absorption spectra of transition metal compounds', F. M. F. de Groot, Z. W. Hu, M. F. Lopez, G. Kaindl, F. Guillot, M. Tronc, J. Chem. Phys., 1994, **101**, 6570–6576.

62 'Multiplet effects in the Ru $L_{2,3}$ x-ray absorption spectra of Ru(IV) and Ru(V) compounds', Z. Hu, H. von Lips, M. S. Golden, J. Fink, G. Kaindl, F. M. F. de Groot, S. Ebbinghaus, A. Reller, Phys. Rev. B, 2000, **61**, 5262–5266.

63 a'x-ray circular dichroism as a probe for orbital magnetization', B. T. Thole, P. Carra, F. Sette, G. van der Laan, Phys. Rev. Lett., 1992, **68**, 1943–1946;b'Orbital-magnetization sum rule for x-ray circular dichroism: A simple proof', M. Altarelli, Phys. Rev. B, 1993, **47**, 597–598.

64 'X-ray circular dichroism and local magnetic fields', P. Carra, B. T. Thole, M. Altarelli, X. Wang, Phys. Rev. Lett., 1993, **70**, 695–697.

65 'Determination of spin- and orbital-moment anisotropies in transition metals by angle-dependent x-ray Magnetic Circular Dichroism', J. Stöhr, H. König, Phys. Rev. Lett., 1995, **75**, 3748–3751.

66 'Magnetic properties of transition-metal multilayers studied with x-ray magnetic circular dichroism spectroscopy', J. Stöhr, R. Nakajima, IBM J. Res. Develop., 1998, **42**, 73–88.

67 'X-ray magnetic circular dichroism – a high energy probe of magnetic properties', T. Funk, A. Deb, S. J. George, H. Wang, S. P. Cramer', Coord. Chem. Rev., 2005, **249**, 3–30.

68 'Integrated x-ray L absorption spectra. Counting holes on Ni complexes. H. Wang, P. Ge, C. G. Riordan, S. Brooker, C. G. Woomer, T. Collins, C. A. Melendres, O. Graudejus, N. Bartlett, S. P. Cramer, J. Phys. Chem. B, 1998, **102**, 8343–8346.

69 'In search of metal hydrides: an x-ray absorption and emission study of [FeNi] hydrogenase model complexes', S. Hugenbruch, H. S. Shafaat, T. Krämer, M. U. Delgado-Jaime, K. Weber, F. Neese, W. Lubitz, S. DeBeer, Phys. Chem. Chem. Phys., 2016, **18**, 10688–10699.

70 'High energy resolution core-level x-ray spectroscopy for electronic and structural characterization of osmium compounds', K A. Lomachenko, C. Garino, E. Gallo, D. Gianolio, R. Gobetto, P. Glatzel, N. Smolentsev,

G. Smolentsev, A. V. Soldatov, C. Lamberti, L. Salassa, Phys. Chem. Chem. Phys., 2013, **15**, 16152–16159.

71 'X-ray absorption near-edge structure calculations beyond the muffin-tin approximation', Y. Joly, Phys. Rev. B, 2001, **63**, 125120.

72 'Self-consistent aspects of x-ray absorption calculations', O. Bunău, Y. Joly, J. Phys.: Condens. Matter, 2009, **21**, 345501.

73 'Resonant inelastic x-ray scattering of molybdenum oxides and sulfides', R. Thomas, J. Kas, P. Glatzel, M. Al Samarai, F. M. F. de Groot, R. Alonso Mori, M. Kavčič, M. Zitnik, K. Bucar, J. J. Rehr, M Tromp, J. Phys. Chem. C, 2015, **119**, 2419– 2426.

74 'The CTM4XAS program for EELS and XAS spectral shape analysis of transition metal *L* edges', E. Stavitski, F. M. F. de Groot, Micron, 2010, **41**, 687–694.

6

Case Studies

Having developed the experimental and analysis basis for XAFS spectroscopy, we show how these techniques can contribute to the understanding of complex and changing materials. The aim is provide a basis for researchers to choose when and how to apply a particular approach to their own materials problems. The strengths of XAFS reside in the following factors:

- Structural information is available for disordered as well as ordered materials.
- XANES features are relatively intense and are sensitive to oxidation state and coordination changes, and this can allow chemical state imaging. XANES features may also provide detail about materials properties as well as structure.
- Scattering of strongly bonded neighbors provides relatively intense features and the bond lengths to them can be determined to 0.02 Å or better.
- Interatomic distances from an absorbing atom to high atomic number elements can also be estimated to about 1% accuracy.
- Angular information can also be derived from multiple scattering and triangulation using two or more absorbing elements within scattering distances of each other.

But this will still provide a partial description of a local structure. Unfortunately, the correlations between intensity related factors like amplitude and Debye-Waller factors mitigate against accurate estimation of coordination numbers, and the normally accepted degree of precision (10–15%) may not discriminate between some structural models. This discrimination can be aided by carrying out simultaneous measurements with complementary spectroscopies, such as infra red, uv-visible, and Raman, and/or by adding wide or small-angle x-ray scattering measurements to the XAFS spectra. Hence examples will show how results from such techniques have contributed to the total study.

6.1   Chemical Processing

A heuristic division for sampling systems is between solution and solid-state chemistry. Processing of bulk and fine chemicals can involve homogeneous catalysts, with all reagents within the single, generally liquid, phase. Alternatively, the reagents in a mobile phase may visit a catalyst in a separate phase, most commonly a solid, often porous (heterogeneous catalysis). Studies of catalysts can also take various forms. The simplest is an *ex situ* study of a catalyst precursor, or, perhaps a recovered material after the catalytic process. In this approach the sample may be presented as any other material and the measurements can be optimized with little compromise. Such measurements are valuable starting points since they can be provide high-quality XAFS spectra of a material of known structure. XAFS analysis of this can give a good expectation of what structural information could be attainable for this study.

However, catalytic reactions may require promoters to generate the active species *in situ*, and the reagents themselves may often be necessary to effect catalyst generation. Hence, the *ex situ* approach alone may shed little or no light on the catalytic process. *In situ*, or *operando*, experiments are essential for circumventing this. These require considerable planning to provide safe remote operation and triggering from a control cabin to an experiment within the x-ray hutch. The gain is a deeper understanding of the structural changes that may accompany the stages in the formation, operation, and deactivation of a catalyst.

However, the *operando* approach by itself may also define the slowest reacting species within a catalytic cycle, and even merely sequestered forms of it. Additional pulsed, non-equilibrium sampling methods can be employed by switching in reagents or varying their concentration, or applying a pulse thermally, electrochemically, or photochemically. These perturbation methods require more demanding experimental design, up to synchronization to the electron bunches of a storage ring, but they can provide a way to identify some of the key steps in a catalytic cycle.

6.1.1   Liquid Phase Reactions

Liquid phase samples have the advantages of generally being homogeneous in nature. Pin-hole effects are absent, and x-ray scattering is relatively weak. Concentrations and path lengths can be defined accurately and so the sample presentation should be well understood. The absorption of an organic solvent, ethanol is presented in Figure 6.1. It can be seen that for the majority of the transition elements, path lengths of several mm are viable, with a few cm feasible for the K edge studies of $4d$ elements. Considering the K edges, only for the lighter elements does solvent absorption become problematical. For example,

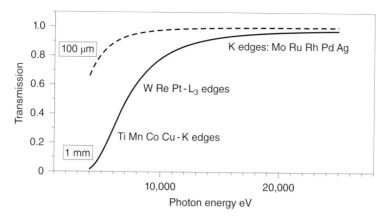

Figure 6.1 X-ray transmission of two path lengths of ethanol, with energies of selected *3d*, *4d*, and *5d* absorption edges.

the attenuation length of ethanol drops from 456 μm for Ti (4966 eV), through 243 μm for Ca (4039 eV) to 60 μm for S (2472 eV). For the light elements too choosing a window material requires careful consideration of their absorption properties, in addition to other physical and chemical characteristics. Sulfur does have a higher linear absorption coefficient (by 44%) at the *K* edge than titanium, but this does not offset the reduction in sampled volume caused by the much lower attenuation length.

For dilute or thin samples, fluorescence detection will become the preferred detection method, but the threshold for this is strongly dependent upon the experimental arrangement and the characteristics of the detectors. The attenuation lengths of the x-ray emissions may become a limiting factor since they may be considerable shorter than at the higher energy of the absorption edge. Taking the example above of three selected elements, the attenuation length of ethanol at the *Kα* energies is reduced to 339 μm for Ti (~4505 eV), 185 μm for Ca (~3690 eV), and 47 μm for S (~2308 eV).

6.1.1.1 Steady State or Slow Reactions (Minutes-Hours)

A simple cell for observing slow reactions is shown in Figure 6.2. Syringe techniques can be employed to introduce air-sensitive solutions and the cell may then electrically heated. In this version, the path length between Kapton windows was fixed at 3 mm. The beveled circular hollow in the supporting plate is to aid fluorescence detection, since the cell itself will restrict the collectable solid angle.

Cell design is governed by the reaction conditions and the x-ray transmission of suitable materials. Using capillaries as sample housings reactions can be monitored under high gas pressures. An example is the hydroformylation of

Figure 6.2 A heatable cells with inlets for air sensitive solutions.

octe-1-ene to nonanal with β-diketonato complex [Rh(acac)(CO)$_2$] as catalyst precursor {acac = CH$_3$C(O)CHC(O)CH$_3$$^-$} in super-critical CO$_2$.[1] At the Rh *K* edge (23.2 keV) x-ray absorption effects are relatively low. The elevated pressure (CO$_2$ 120 bar, H$_2$ 10 bar, CO 10 bar) was contained in a capillary bored through a PEEK rod. Heating via a hot air blower avoided cell constraints on the fluorescence collection angle, and so data could be measured in under 4 h for these dilute solutions in scCO$_2$ using a 13-element Ge detector (ID26, ESRF), as shown in Figure 6.3.

The Rh XAFS data showed a progression of reaction changes to products stable over the acquisition time. At room temperature, one carbonyl group was replaced by a phosphine ligand (Rh-P 2.25 Å). With a PEt$_3$:Rh ratio of 1:3, the coordination site could be shown by EXAFS to contain 3 phosphine and 1 CO ligand at 43 °C, the β-diketonato ligand having been removed under the reaction conditions. Only by comparison with previous IR studies could the formula of [RhH(CO)(PEt$_3$)$_3$] be proposed (Figure 6.4). With a PEt$_3$:Rh ratio of 1:1 and under hydroformylation conditions, the $k^3.\chi(k)$ data extended to a higher *k* value, suggesting the presence of a heavier back-scatterer, which could be refined as Rh-Rh (2.70 Å). Hence under these circumstances rhodium was being sequestered as dimers or higher clusters, rather than as the metal hydride required for the catalytic cycle for hydroformylation.

6.1.1.2 Fast Reactions (Ms to Minutes)

In the example above, there were clearly a number of chemical steps that were not observed, for example, during the loss of the chelating β-diketonato ligand.

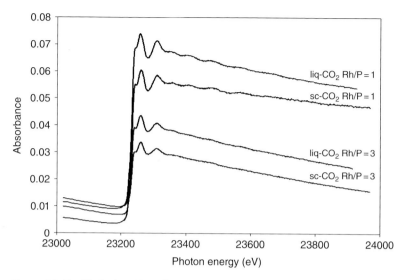

Figure 6.3 The Rh *K* edge XAFS of 2–3 mM [Rh(acac)(CO)$_2$]/PEt$_3$ in a mixture of oct-1-ene, hydrogen, carbon monoxide in (c) l-CO$_2$ and scCO$_2$ at Rh/P = 1 and 3 (*Source:* Fiddy (2004).[1] Reproduced with permission of Royal Society of Chemistry).

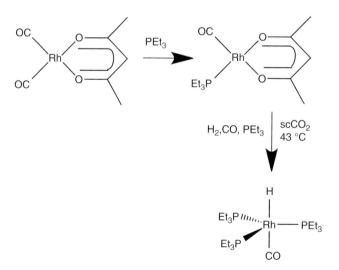

Figure 6.4 Reaction scheme for the activation of a rhodium catalyst for the hydroformylation of octene in scCO$_2$.

For the EXAFS data to be valid as is, the observed species must outlive the acquisition time, and that minimizes significantly the application of *operando* approaches for probing active, dilute catalysts. One straightforward way to circumvent this is to transfer an aliquot into an observation tube and then

Figure 6.5 Arrangement for observing a frozen solution by fluorescence detection. Cooling is by an Oxford Instruments Cryostream with an outlet temperature of 100 K (*Source:* B18, Diamond).

quench the reaction with a cold gas stream[2] (Figure 6.5). This proved valuable during a study of chromium catalysts for selective oligomerization of ethene, where the combination of chromium concentration and low attenuation length meant that lengthy acquisition times were required.[3,4]

Stopped-flow techniques provide a means of mixing solutions within a few ms, and thus provide a means of tracking a reaction process from that initial timescale. It is thus a valuable method for investigating the first few minutes of a reaction. There are four general means of monitoring the reactions of these solutions by XAFS.

1) The first method was by energy dispersive EXAFS.[5] In this way the entire spectrum is recorded simultaneously, but the prime restriction is that it requires transmission detection. If the chemical system satisfies this criterion, then this will provide the fastest experiments in real time to complete a suite of kinetic runs.

2) The second method involved time scans at fixed energies, repeating the process 500 times to allow a time-resolved set of XAFS spectra to be generated.[6] This process can be carried out both in transmission and fluorescence, although in transmission it will take much more elapsed time that a dispersive measurement.

3) By combining the rapid mixing with a freeze quench, then the time resolution of the stopped-flow technique is maintained and for dilute samples, this preserves the state of the sample during a lengthy data acquisition of a rapidly mixed solution.[2,7]

4) Stopped flow may also be combined with QEXAFS scanning, which also allows both transmission and fluorescence detection.[7] Unlike all of the other methods, there is a risk of the spectrum changing during the duration of a scan, meaning that the observed spectrum is due to different mixtures of species across the scanned energy. This can be minimized by very rapid scanning instrumentation,[8] but this reduces the acquisition time per data point, and this can result in much lower photon counts per pixel as compared to a dispersive experiment.

The balance between these options can be clear in many cases, but for many studies it will be availability that governs the choice. Stopped-flow mixers can be fouled by formation of precipitates; this makes experimentation, at best, frustrating. It has been shown that microfluidic methods can cope with this circumstance. The XAFS measurement is essentially static in this configuration, and the time resolution comes from the distance travelling along the flow tube after mixing.[9] The small beams available from modern light sources match microfluidic systems well and can provide sub-second time resolutions.

In Figure 6.6 we show a stopped-flow cell with the alignment laser illuminating the direction of the transmission XAFS. There is also an optical spectroscopy sampling system perpendicular to the x-ray beam axis for simultaneous monitoring. Earlier, activation of nickel homogeneous catalysts by aluminum alkyls (for alkene oligomerization) had been monitored by uv-visible externally to energy dispersive XAFS spectroscopy (Station 9.3 at the SRS).[10] Whilst this can provide kinetic evidence that can be cross-referenced between the two techniques, validation is more clear cut if the techniques are carried

Figure 6.6 A stopped-flow cell (Biologic) for transmission XAFS and simultaneous uv-visible spectroscopy mediated by optical fibers (*Source:* B18, Diamond).

out simultaneously, as can be achieved using this combined spectroscopic arrangement.

This combination proved to be invaluable during a study of the reduction of $[Ir^{IV}Cl_6]^{2-}$ by $[Co^{II}(CN)_5]^{3-}$ by energy-dispersive EXAFS using the Ir L_3 absorption edge.[11] In all the stages the iridium site has 6 chlorides as neighbors. Accompanying the oxidation state reduction is a change from a t_{2g}^5 to a t_{2g}^6 electron configuration in these octahedral complexes (Figure 6.7) This addition of a non-bonding electron has only a small effect on the Ir-Cl bondlengths (0.03 Å). Hence the uv-visible spectroscopy helped indicate the stages that were being monitored by XAFS. It became clear that within 200 ms the electron transfer the bridged intermediate had been formed and the electron transfer through the bridge had occurred. The dissociation of the dimer, predominately via Co-Cl bond cleavage, was being observed, with a rate of 0.6 ± 0.002 s^{-1}. The quality of the data from these reaction solutions (80 mM), however, allowed the reaction to be monitored (Figure 6.8) and, using the methods described in Section 5.2.2, product and intermediate could be analyzed.

The electron transfer system was monitored by a single wavelength uv-visible spectrometer, which can provide very accurate absorbance measurements. However, this meant that the whole XAFS/uv-visible experiment was carried out several times at key wavelengths for the optical spectrometer. An alternative approach is to match the characteristics of the XAFS measurement using dispersion optics also in a diode-array uv-visible spectrometer.

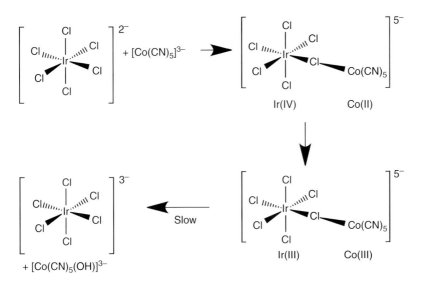

Figure 6.7 Reaction scheme for the inner-sphere electron transfer reaction between $[IrCl_6]^{2-}$ and $[Co(CN)_5]^{3-}$.

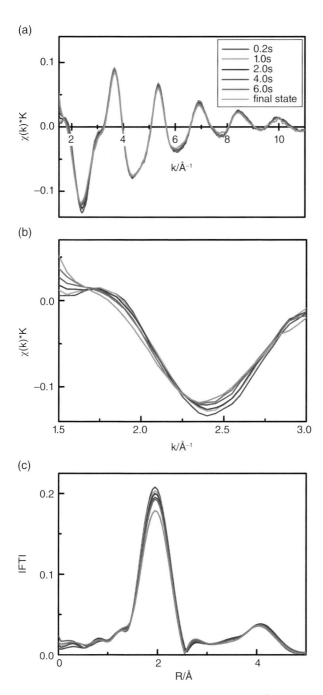

Figure 6.8 Ir L_3 edge EXAFS from the reaction of $[IrCl_6]^{2-}$ with $[Co(CN)_5]^{3-}$ (80 mM) using a stopped-flow system (ID24, ESRF). a) EXAFS of the time series, b) expansion of a), c) Fourier transforms (*Source:* Diaz-Moreno (2005).[11] Reproduced with permission of Royal Society of Chemistry).

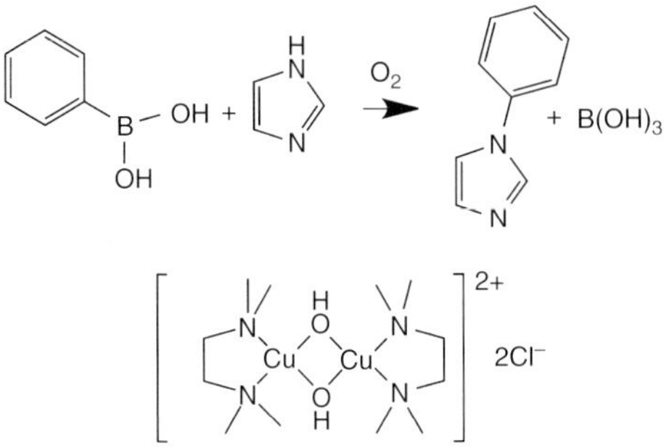

Figure 6.9 Arylation of imidazole by phenylboronic acid with the copper catalyst.

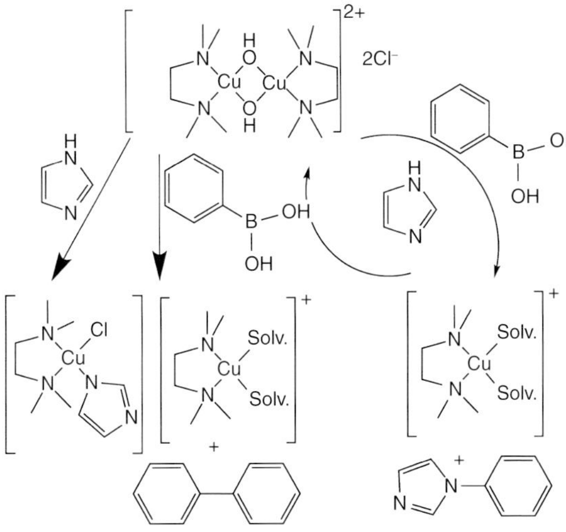

Figure 6.10 Observed reactions between the Cu catalyst in Figure 6.9 with the catalysis reagents in aqueous NMP (N-methylpyrrolidone).

This combination was one aspect of a wider study of the arylation of imidazole, which is catalyzed by a Cu(II) complex of TMEDA (tetramethylethylenediamine), as shown in Figure 6.9.[12]

The two reagents can be added stepwise so that steps in the catalytic cycle can investigated as discrete events. Very small but clear time-resolved (every 3 seconds) changes in the XANES of the Cu K edge and the uv-visible spectra were observed on addition of imidazole. Taken in conjunction with other evidence, this indicated maintaining a Cu(II) site, now mononuclear, with coordinated imidazole and chloride (Figure 6.10). The alternative stoichiometric addition of phenylboronic acid results in clear changes in Cu XANES, with the growth in a pre-edge peak attributable to a *1s-4p* transition for Cu(I) ($3d^{10}$);

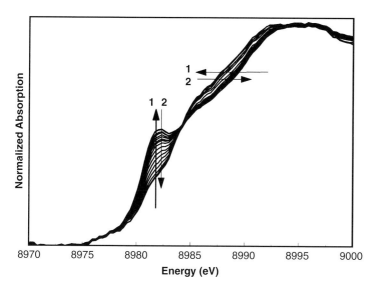

Figure 6.11 Energy dispersive Cu *K* edge XAFS of the reaction: [Cu(OH)(TMEDA)]$_2$Cl$_2$], imidazole and PhB(OH)$_2$. Maximum after ~30 s (arrow 1), after which the features decrease again (arrow 2) (*Source:* Tromp (2010).[12] Reproduced with permission of American Chemical Society).

this is supported by the loss of the *d-d* band in the visible region (625 nm). A plausible mechanism would involve an oxidative phenylation of the copper to transient Cu(III), followed by a reductive elimination to generate biphenyl leaving a Cu(I) behind. When both reagents are present, then a reversible change in the copper oxidation state is observed (Figure 6.11). In this case the coupling forms N-phenylimidazole, so a Cu(III) transient with both radicals attached was proposed. The reoxidation of the Cu(I) centers appears to be mediated by phenylboronic acid, and allows the completion of the catalytic cycle (Figure 6.10).

6.1.1.3   Very Fast Reactions (~100 ps – ms)

All experiments described so far have treated the storage ring as a continuous source, but as acquisition times are reduced toward the μs range this can no longer be assumed (Section 3.1.4). Electron bunches travel ~ 300 m in 1 μs. In a multibunch filling there will be several 100 bunches in a storage ring, so in a medium-sized storage ring there are tens of thousands of bunches passing per ms, giving an effectively continuous source. Storage ring–filling modes are not always evenly spaced, and this unevenness is detectable certainly in the tens of μs time range. EXAFS in particular is very sensitive to small intensity changes. This problem can be avoided if the XAFS experiment is synchronized to the rotation of the electron beam and so the same intensity pattern can be utilized for each measurement. For timescales below the ms, such synchronization to the machine clock should be considered. The alternative view is that over several hours of acquisition everything averages out.

Some storage ring filling modes are developed for time-resolved studies in particular. The simplest is a having a single bunch injected. This would normally have an enhanced current over a bunch in a multibunch filling pattern, but the current will be reduced by a factor of about 20–30. Provided the detection method can be windowed within the visits by the bunch, the potential time resolution will become governed by the bunch length as it traverses the observation point. Generally, this is in the region of 10 to 100 ps.

In a storage ring of ~ 300 m circumference a synchronized experiment operating with single bunch filling at a repetition rate of 1 MHz might receive about 5% of the photons that a standard multibunch would receive in the same time period. Assuming the normal square relationship to recover single/noise, the elapsed time to achieve this would be increased by 400, or, a 10-minute multibunch experiment requires about 3 days. This can be reduced if the storage ring were operating in a few-bunch (e.g., 4) mode, providing that the reduced gap between pulses can be accommodated by the characteristics of the detector and experimental arrangement. Such modes do not receive universal acclaim from the majority of SR users, and thus do not receive high priority in scheduling. A more common compromise is a hybrid mode, which may have one or two larger bunches a few 100 ns gap to a multibunch filling region. Achieving a 1-MHz repetition rate for any experiment is not easy, and the repetition rate of an exciting pulse or sample refreshing may reduce this perhaps by 10^3 to 10^5. A special case is provided by the APS. The standard operation mode provides 24 bunches of nominal current 4.25 mA spaced by 153 ns, and is well suited for time-resolved studies in the 100 ps to ns regime.

Considering the options for pulsed experiments in Section 6.1.1.2, two of the four are barely feasible for these much shorter timescales: QEXAFS and freeze-quench methods would be very demanding on a (sub) μs timescale. The prevalent method is the step-scan approach building up time series data at each x-ray energy; the dispersive approach has also been demonstrated with the advantage of deriving all of the x-ray spectrum simultaneously, albeit with very restricted options for detection [13]. Flow experiments were also developed to investigate the structures of excited states with μs time resolution.[14]

XAFS was established as a means of investigation photo-excited states particularly for a series of metal-to-ligand charge transfer states (Figure 6.12). In the example of the $3d^{10}$ complex [Cu^I(dmp)$_2$]$^+$ (dmp = 2,9-dimethyl-1,10-phenanthroline), which displays a strong MLCT band at 450 nm, the observable excited state displays XANES features almost identical to those of the Cu(II) analogue,[15] as shown in Figure 6.13. The Cu(II) center with a $3d^9$ configuration is expected to undergo a Jahn-Teller distortion. DFT and XANES calculations have been used to discriminate between an angular distortion resulting in

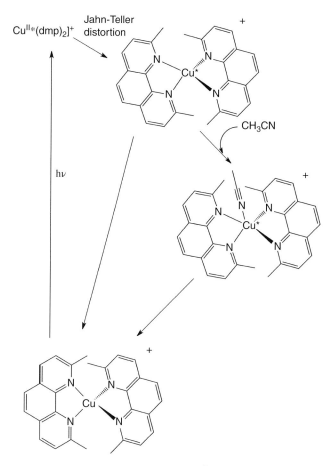

$Cu^{II}*(dmp)_2]^+$ Jahn-Teller distortion

$h\nu$

$- CH_3CN$

Figure 6.12 Photoreaction scheme for $[Cu^I(dmp)_2]^+$ in acetonitrile solution. Laser excitation creates transient MLCT excited state in which the Cu^{II} excited state that flattens and may also form an exciplex with a solvent molecule.

a flattened four-coordinate and exciplex model for the observable excited state. The time resolution of the experiment was limited by the single bunch length, which provided a 100 ps x-ray pulse.

This experiment was carried out on a 2 mM solution with a 0.5-mm path length using a multi-element Ge detector to monitor the fluorescence signal in a time-series, x-ray step scan acquisition mode. The laser repetition rate was 1 kHz, much lower than the rotation rate of the single bunch used (271 kHz), and the single bunch current (5 mA) was 5% of the total ring current, and thus only 0.018% of the available x-ray light was utilized, giving a total acquisition time of 40 hours. Nevertheless, EXAFS data was obtained to about 400 eV after the edge ($k \approx 10$ Å^{-1}), and this could be used to estimate the Cu-N distance(s) in

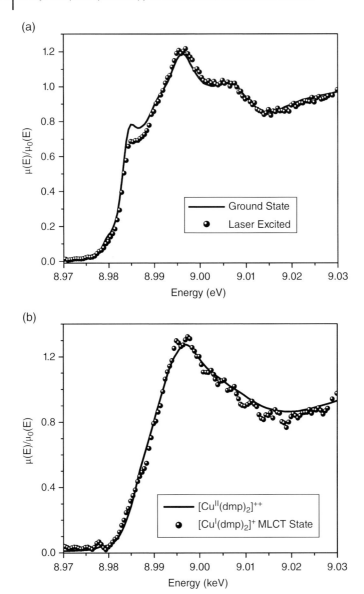

Figure 6.13 The Cu *K* edge XANES of the ground state and laser excited [Cu^I(dmp)$_2$]PF$_6$ (2 mM) in acetonitrile solution. Irradiation via a Nd/LF laser, $\lambda = 527$ nm, 1 mJ/pulse, 5 ps FWHM (APS, 11ID-D) (*Source:* Chen (2003).[15] Reproduced with permission of American Chemical Society).

the exciplex, possibly also differentiating between the shorter distance to the acetonitrile ligand (2.03 compared to 2.07 Å). To establish these results, the XAFS spectra of the excited state (with 20% conversion) had to extracted from the majority, ground state contribution. More recently, using a detection system optimized to the bunch structure, the total experimental time has been reduced significantly,[16] although it is still in the region of 20 h.

Improving on this acquisition time can, in principle, be achieved by increasing the incoming x-ray flux on the sample. But on a given beamline, there are the following options.

1) The photo-conversion could be increased by a higher laser power. In practice though, it is hard even to reach the 20% conversion in this example, given the difficulties in dealing with the laser power on the sample. Balancing laser power, and sample absorption in the uv-visible and x-ray regions requires care.

2) A factor of a few 100 can be gained by using an energy dispersive geometry, albeit with the concentration necessary for a transmission measurement [13].

3) Alternate light-on, light-off measurements should be made in fast succession to provide the difference spectrum with minimum beam movement.

4) The laser repetition rate can be increased, thus sampling a proportionate increase in the available x-ray pulses, and the sample refreshed by continuous flow.

This approach was well-developed at the Swiss Light Source.[17] In a hybrid mode, a single bunch provided an x-ray pulse (85 ps FWHM) every 960 ns. The laser retention rate was adjusted so that every such pulse could be utilized on a light-on-light-off alternation, giving an experiment repetition rate of 520 kHz. Compared to a 1 kHz repletion rate this could afford an increase in S/N of $\sim$ 23, and acquisition time of about 1 hour.

6.1.1.4 Ultrafast Reactions (fs – ps)

In this time regime, only a fraction of a normal single bunch can be sampled. For example, a streak camera detector can be used to provide a time structure with the bunch. The length of single bunch itself can be reduced by operating the storage ring in a particular mode (termed low α), with a low current. For example Diamond can provide bunch lengths in the range of 3–5 ps, with currents of a few 10s of μA, about $2\text{-}3 \times 10^{-3}$ of a normal single bunch. Alternatively, a slice of a bunch can be deflected by a laser and that slice be used as the x-ray source. In this way a pulse duration of 140 ± 30 fs has been generated from a 4 mA, 85 ps single bunch;[18] (about 10^5 photons/sec (0.1% bandwidth, in the energy range 5–8 keV) at 1 kHz ($\sim$100 photon/pulse) arise from this source into the beamline.

However, the most dramatic gain is to acquire spectra with a time resolution similar to some molecular vibrations. A metal-ligand vibration at 300 cm^{-1} will

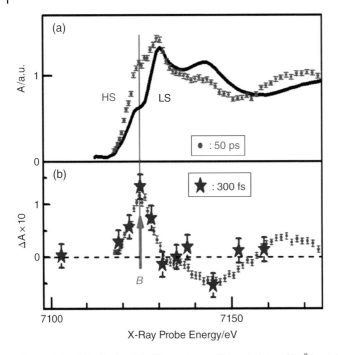

Figure 6.14 a) Fe K edge XANES spectrum of the LS state of [Fe^{II}(bpy)$_3$]$^{2+}$ (circles) and of the HS state at 50 ps time. (b) Transient XANES spectrum at 50 ps (red dots) and at 300 fs (blue stars) time delays (*Source:* Cannizzo (2012).[19] Reproduced with permission of Elsevier).

have a time constant of 111 fs, so transients may be observable that show little change after the initial excitation. As an example, the excitation of the low spin d^6 ($^1A_1$) complex, [Fe(bipy)$_3$]$^{2+}$ (bipy = 2,2'-bipyridyl) creates the high spin excited state ($^5T_2$) present 50 ps after the laser pulse. Accessing this state involves a two-electron movement and is doubly spin forbidden. As the information shown in Figure 6.14 shows, the cascade for a ^{1}MLCT excited state to create the ligand-field excited state had already occurred within 300 fs.[19]

Still faster experiments are in the realm of free electron lasers that provide similar photon counts within a pulse as are provided per second in a storage ring source (Section 4.6). Examples of early experiments at LCLS have been outlined in Section 4.6.4. These are natural extensions of the approach, including studies of the photo-physics and -chemistry of [Fe(bipy)$_3$]$^{2+}$,[20] [Ni(tetramesitylporphyrin)] [21] and photosystem II [22].

6.1.2 Reactions of Solid-State Materials

Many of the factors influencing XAFS measurements are similar for solid-state materials. The main differences are in the care required for:

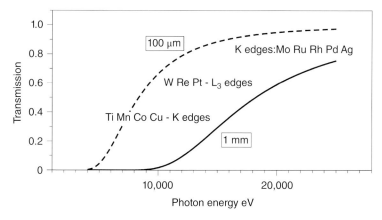

Figure 6.15 X-ray transmission of two path lengths of SiO$_2$ ($\rho = 2.27$ gcm^{-3}), with energies of 3d, 4d, and 5d absorption edges.

1) presentation of powdered materials, which are intrinsically textured, to avoid pin-holes in the sample;
2) potential complication of x-ray polarization for oriented or single crystal samples and
3) likely background x-ray absorption from the bulk of the material.

The last factor is illustrated with the curves in Figure 6.15, with SiO$_2$ used as a common base material in catalysis and geochemistry. For un-pressed powdered material, the effective density may be lower by a factor of ∼ 2, which would lower the linear absorption coefficient proportionately, but it is apparent that sample absorption requires significant consideration for solid samples. At any energy below 10 keV, a 1-mm sample thickness will absorb extremely strongly reducing the efficacy of transmission measurements; the same conclusion applies below the titanium *K* edge for a 100-μm sample thickness.

At lower energies, where there is a high density of accessible absorption edges, a film of 10 μm will afford viable transmission down to the phosphorus *K* edge, and then, below the Si *K* edge for Al and Mg (Figure 6.16). However, at such a sample thickness the absorption jumps will be small, and alternative detection methods may provide better quality spectra.

6.1.2.1 Steady-State or Slow Reactions (Minutes-Hours)

An example of a soft x-ray absorption study recorded in transmission has utilized this window below the Si *K* edge for an aluminosilicate.[23] The de-alumination of NH$_4$-Y zeolite by steaming was investigated using the Al *K* edge. Even at the same oxidation state (Al^{III}), there is a significant difference in edge position as well as XANES shape between octahedral and tetrahedral sites (Figure 6.17). The XANES spectrum of NH$_4$-Y is typical for the

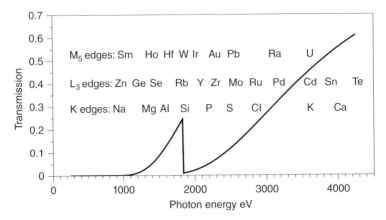

Figure 6.16 X-ray transmission of SiO_2 ($\rho = 2.27$ gcm^{-3}) of 10 μm path length, with energies of selected K, L_3, and M_5 absorption edges (1–4.5 keV).

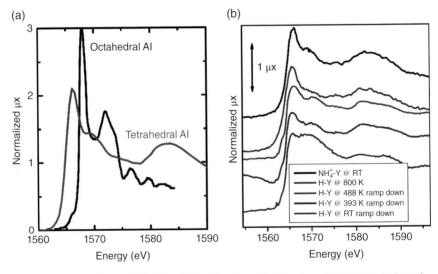

Figure 6.17 Al K edge XANES of a) α-Al_2O_3 (black) and Na-Y (red), and b) NH_4-Y (black) and H-Y zeolites during heating and cooling in a steamed de-alumination (*Source:* Agostini (2010).[23] Reproduced with permission of American Chemical Society).

tetrahedral sites in the zeolite, but on cooling there are additional features to high energy of this absorption edge, in a region expected for an octahedral site. The formation of six-coordinate aluminum was confirmed by ^{27}Al NMR, which also identified five-coordinate sites being developed during de-alumination. Careful analysis of synchrotron powder diffraction results demonstrated the formation of extra framework Al sites during the heating process.

6.1.2.2 Fast Reactions (Ms to Minutes)

At higher x-ray energies, the constraints on mounting reactors are much reduced and the additional flexibility provides options for simultaneous measurements. For example, the air heater in Figure 6.18 allows the quartz reactor tube to be investigated by transmission, fluorescence and emission,[24] x-ray diffraction, as well as optical methods.

Although restricted to transmission XAFS, the cell system in Figure 6.19 allows simultaneous measurement of the IR spectrum of a powdered catalyzed contained in a boron nitride walled cell under a controlled ambient gas mixture. With a rapid scanning IR spectrometer and energy dispersive XAFS on ID24 at the ESRF, then repetition rates of 10 Hz or higher were feasible. In these circumstances the resolution of the experiment becomes limited by the properties of the gas pulse passing through the cell. In these ways the structural changes that occur at a catalyst can be monitored as a function of temperature and gas composition.[25,26] The x-ray beam was sited just under the surface of the catalyst bed, with the IR beam sampling the surface of the bed. In this way, XAFS spectra were recorded on a slice of the region sampled by IR. Experiments at the *4d K* edges presented a favorable option since the x-ray transmission was reasonably high for a 4- to 5-mm bed necessary for good IR sensitivity.

An example of the results of this approach for a 4 wt% Rh catalyst, activated *in situ* and on a non-porous γ-alumina support, are given in Figure 6.20.[27] Switching in the O_2 into a helium stream caused a high degree of oxidation of the Rh, which was largely reversed on switching the flow to CO. This reduction was accompanied a pulse of CO_2 and then IR bands for linear and bridging

Figure 6.18 *In situ* reactor cell for gas-solid reactions, including heterogeneous catalysis. The temperature control uses a hot-air blower mediated by an *in situ* thermocouple. Gas composition is controlled by switching valves and mass flow controllers, and the output analyzed by mass spectrometry (multiple ion monitoring). The window for a nine-element Ge fluorescence detector is in view (*Source:* Diamond, B18).

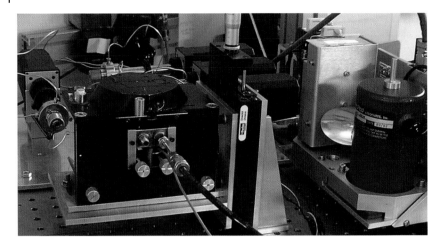

Figure 6.19 Combined diffuse reflectance Fourier transform infrared spectroscopy (DRIFTS) and transmission XAFS cell, showing IR detector. Gas composition is controlled by switching valves and lass flow controllers, and the output analyzed by mass spectrometry (multiple ion monitoring) (*Source:* ESRF, ID24).

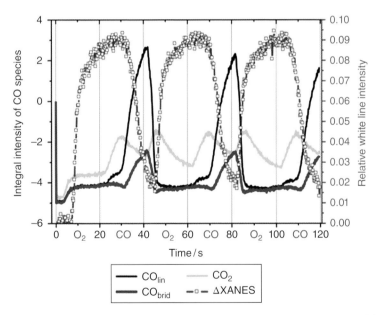

Figure 6.20 Gas switching between 5% O_2/He and 5% CO/He at 573 *K*. Plot of changes in the Rh *K* edge XANES intensity at 23250 eV, CO_2 concentration, and the IR bands of linear and bridged CO sites on 4 wt% Rh/γ-Al_2O_3 (*Source:* Kroner, http://www.tandfonline.com/action/showCopyRight?scroll=top&doi=10.1080%2F2055074X.2016.1266762. Used under CC-By4.0 https://creativecommons.org/licenses/by/4.0/).

carbonyl on metallic rhodium could be observed. These bands are rapidly lost when O_2 is reintroduced, and they are converted into CO_2. EXAFS analysis shows that these redox processes are accompanied by the making and breaking of metal-metal bonds, changing the mean particle size of the metallic fraction.

Enhancing the time resolution for gas-solid reactions can become problematical from a sample viewpoint. Single-turnover mapping of porous catalysts in particular show very strong diffusional effects.[28] As a result, the active sites within micro- and meso-porous materials, and in lamellar clays may be restricted to the periphery of particles. But in many XAFS measurements the entire bulk of the particle is sampled, and thus may not be representative of the site turnover the catalysis. For gas-solid reactions, the best time resolution will be derived from flat surfaces, with molecular beams as the gas source and/or laser-excitation.

Imaging by XAFS has also demonstrated texture in the state of the metal across a catalyst bed. For example, the state of platinum in a 2.5 wt% Rh- 2.5 wt% Pt/Al_2O_3 has been imaged using the difference in the white line height of the Pt L_3 edge between oxidized and reduced platinum under methane oxidation conditions with an excess of methane.[29] In the center of the catalyst bed a progression of reduced platinum is generated along the bed as the temperature is raised through the light-off temperature for the combustion as the atmosphere reaching that part of the bed becomes more reducing with the removal of O_2 by the combustion (Figure 6.21).

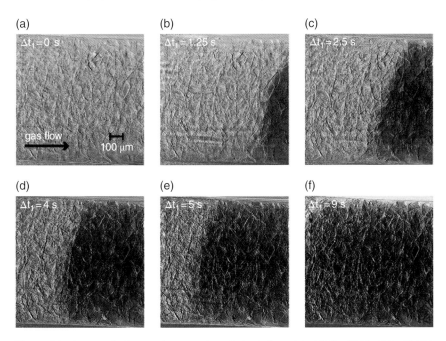

Figure 6.21 Image of a 1-mm x 1-mm region of a bed of a catalyst (2.5 wt% Rh- 2.5 wt% Pt/Al_2O_3) using x-rays at 11596 eV. The oxidized (light) and reduced (dark) regions are distinguished by x-ray transmission at the Pt L_3 white line (*Source:* Kimmerle (2009).[29] Reproduced with permission of American Chemical Society).

6.2 Functional Materials

Materials with opto-electronic or magnetic properties may be investigated in similar ways to those described above, with appropriate environmental control. An additional option is through the investigation of x-ray magnetic dichroism effects, most readily observed in the soft x-ray region using a helical undulator as the light source (Sections 2.3, 4.4.3). X-ray magnetic circular dichroism can be observed on spontaneously magnetized materials (ferro- or ferri-magnets), or when spin polarization is induced into a paramagnetic material by a magnetic field (Section 5.3.2). For example, field-induced x-ray magnetic dichroism effects could be observed for the d^9 Cu^{II}-phthalocyanine molecule (CuPc – a macrocyclic N_4 donor) using the Cu L edges, both as a solid and isolated on a Ag(100) surface.[30] The effect per Cu atom was much lower for the solid sample, due to anti-ferromagnetic coupling. On the silver surface, a monolayer of Cu-Pc forms a square unit cell (14.5 Å). The XAFS spectra were measured in total electron yield, and the tail from the M edges (368–719 eV) of the much more populous silver atoms affected the background of the Cu L_3 (~ 932.7 eV) and L_2 edges (~ 952.3 eV) (Figure 6.22). Using x-ray linear dichroism, the polarization of the spin producing these effects is shown to be parallel with the silver surface, consistent with the odd electron residing in the $3d(x^2-y^2)$ orbital in the tetra-nitrogen coordination plane.

Perhaps more unexpected magnetic behavior has been found in ZnO nanoparticles capped with long-chain organic ligands, since Zn^{II} has a $3d^{10}$

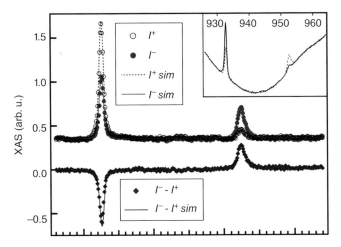

Figure 6.22 Circularly polarized Cu $L_{2,3}$ edges of 1 ML CuPc/Ag(100) at 6 K showing observed spectra with different circular polarization directions, and the difference spectra, with theoretical simulations on the basis of $2p^6 3d^9$ → $2p^5 3d^{10}$ transitions. Inset shows the spectra before background removal (*Source:* Stepanow (2010).[30] Reproduced with permission of American Physical Society).

configuration. However, XMCD effects have been observed at the Zn K edge, indicating that the magnetic signal emanates from the *4p* electrons.[31] A detailed analysis suggests that this comprises two classes of behavior: a paramagnetic signal and a ferromagnetic one formed from about 5–15% of the zinc atoms. This latter component displays saturation and that behavior and varied with the nature of capping ligand (Figure 6.23), indication that the ferromagnetism could be ascribed to the surface region of the nanoparticles.

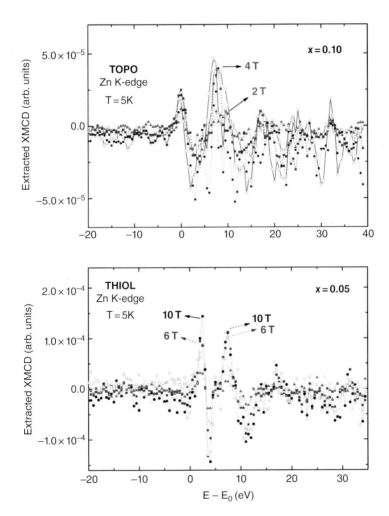

Figure 6.23 XMCD component at the Zn K edge for the ferromagnetic portion of ligand-capped ZnO nanoparticles recorded at 5 K and different applied magnetic fields; Capping ligands: top TOPO {OP(*n*-C_8H_{17})$_3$}; bottom THIOL = ($C_{12}H_{25}$SH) (*Source:* Guglieri (2012).[31] Reproduced with permission of American Chemical Society).

6.3   Imaging on Natural, Environmental, and Heritage Materials

In a similar vein, soft x-ray scanning transmission microscopy (STXM) has been used to investigate the Fe $L_{2,3}$ edge XMCD for magnetosomes in marine magnetotactic bacteria.[32] Particles of 30 nm diameter could be imaged using the difference in transmission between the pre-edge and edge region of the Fe L_3 edge. The XMCD signal of a chain of these magnetosomes was averaged, and shown to compare with that of magnetite itself (Figure 6.24). That the signal was observed without an applied magnetic field implies too that the particles were ferromagnetic.

In a study of iron minerals in metamorphic rocks and clays, mineral speciation was carried out using the Fe K edge XANES, the polarization effects of the x-ray source taken into account, by sample rotation perpendicular with respect to the x-ray beam;[33] in a layered mineral like chlorite the sample deposition often occurs with a preferential orientation of the crystallographic c axis, which is perpendicular to the clay layers. Indeed, the effect of the anisotropy on the XANES features enabled the orientation of the c axis to be mapped within a sample, as well as the oxidation state distribution (Figure 6.25).

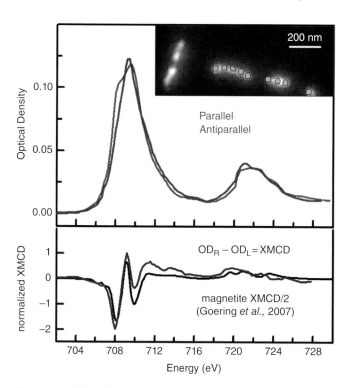

Figure 6.24 Plots of the Fe $L_{2,3}$ edges of 9 magnetosomes (averaged) in magnetotactic vibrio strain MV-1 recorded with the circular polarization parallel and antiparallel to magnetite (*Source:* Lam (2010).[32] Reproduced with permission of Elsevier).

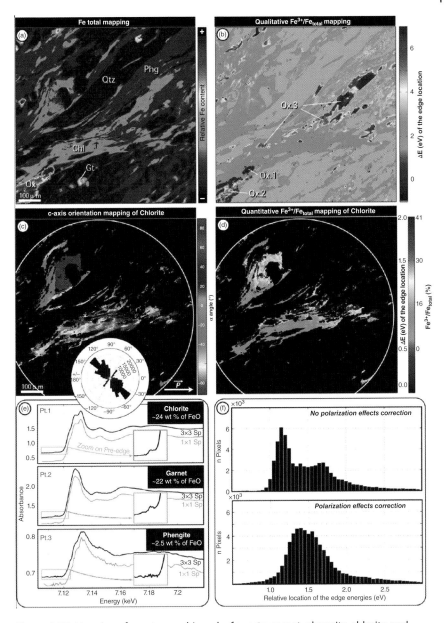

Figure 6.25 Mapping of a metamorphic rock of quartz, garnet, phengite, chlorite, and oxides. a) Total Fe content, b) proportion of Fe(III), c) *c* axis mapping of chlorite, and d) the proportion of Fe(III) in the chlorite (*Source:* Andrade (2011).[33] Reproduced with permission of American Chemical Society).

Chemistry within naturally occurring organic materials has been high-lighted by analysis and conservation efforts for the waterlogged wood historical shipwrecks.[34] A key issue is the oxidation of organosulfur to sulfate (VI), with the attendant risk of material damage form the resulting acid. Speciation of sulfur is impeded by its NMR properties and paucity of uv-visible spectroscopy properties in many chemical states. However, the S K edge XANES varies greatly with oxidation state, and so can be used to differentiate between present and potential sulfate. A section of a plank (0−3 mm from the surface with 3.0% S and 1.4% Fe) from a Danish warship, which sank in 1645 outside Gothenburg, has been analyzed by S K edge XANES (Figure 6.26).

The spectrum shows a group of overlapping peaks near 2473 eV assigned to the sulfur units in thioproteins, and some elemental sulfur. In addition there are prominent peaks due to the white line of S(VI) ($3p^0$) attributable to organic and inorganic sulfate.

The degree of oxidation, unsurprisingly, increases closer to the surface of the wood, as shown in a study of an oak post from the *Mary Rose* (Figure 6.27), which sank in Portsmouth Harbor in 1545. The timbers involved here were studied after a relatively recent retrieval (2005).[35] The combination of prox-imity to the surface and the presence of Fe(III), the occurrence of which does not vary greatly with depth, both strongly correlate with the degree of sulfate formation, thus presenting considerable challenges for conservation.

Pigment samples taken from two ancient (∼18000 BCE) paintings, the "Great Black Bull" and the "bichrome Horse" in the caves at Lascaux (Dordogne, France) and analyzed by XAFS show that different manganese containing minerals,[36] one with Mn(III) and Mn(IV), and the other being MnO_2. It appears that these minerals were used as pigments without much heat treat-ment. Much later historical artifacts containing manganese are found in medieval stained glasses, as in the fourteenth-century glasses in cathedral du Bosc (Normandy, France). Glass protected from weathering displays a Mn K edge XANES typical of Mn(II), but exposure to weather results in the forma-tion of Mn(IV) oxy-hydroxides, with principal component analysis supporting the intermediate samples as being differing mixtures of these two end states. This, and a similar pattern for iron in the silicate glasses, results in a reduction of the luminosity within the cathedral.

A combination of x-ray fluorescence mapping and XANES has been used to reveal the essence of the face of a woman over-painted on a canvass by Vincent van Gogh, which since 1887 is the *Patch of Grass* [37]. The elemental maps show the use of Pb, Hg (vermillion), Sb,and Zn (zinc white) pigments, with the antimony image clear (Figure 6.28). Comparison with the L_3 and L_1 XANES of a series of materials, supports the antimony as being in the pigment Naples

(a)

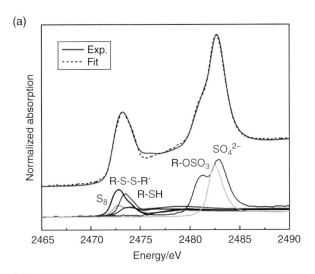

(b)

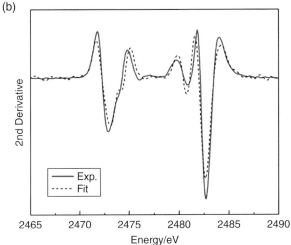

Figure 6.26 S *K* edge XANES of the surface layer of a plank from the *Stora Sofia*, showing the experimental and fitted spectra and second derivative (*Source:* Fors (2011).[34] Courtesy of The Japan Society of Analytical Chemistry).

yellow $[Pb(SbO_3)_2.Pb_3(Sb_3O_4)_2]$, with $Sb(V)$ centers, rather than antimony white (Sb_2O_3). Thus an idea of the color of the original painting can be reconstructed from mixtures of the components identified.

Thus XAFS can be used to probe textured compositions fashioned and then hidden by its talented designer.

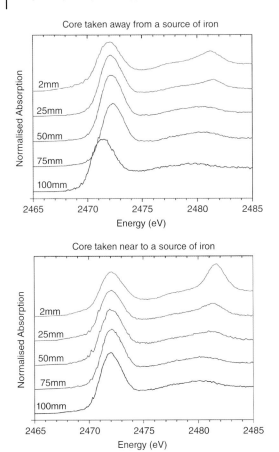

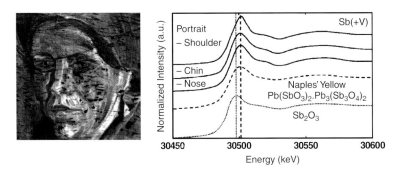

Figure 6.27 S *K* edge XANES as a function of depth from samples of an oak stem post, close to and distance from an original iron fixing (*Source:* Wetherall (2008).[35] Reproduced with permission of Elsevier).

Figure 6.28 Left: Sb *K* emission map of a region of the painting *Patch of Grass* by Vincent van Gogh. Right: Sb *K* edge XANES of regions of the painting compared to those of Naples yellow and antimony white (*Source:* Janssens (2008).[37] Reproduced with permission of American Chemical Society).

6.4 Questions

1 You are considering investigating a homogeneous catalysis process involving a copper(II) complex in a solution of ethanol. It is thought that there may be an active monomeric species, which predominates at a concentration of 1 mM, and is in rapid equilibrium with a less active dimer, predominating at 100 mM [Cu]. Mass absorption coefficients indicate that the x-ray absorption of a 20 mM solution with a path length of 1 mm increase from 0.121 to 0.131 on crossing the Cu K edge.

   A Using Figure 6.1 as a guideline, suggest a good path length for your solution cell for studying the structure of the dimer. What would be the edge jump of this sample?

   B What would be the edge jump of the sample that you would use to study the structure of the monomer? Which detection modes would you consider using for this solution?

2 Consider planning *in situ* experiments on two bimetallic catalysts of composition **A**: 1% wt %Re/1 wt %Pt/SiO$_2$ and **B**: 1% wt %Au/1 wt %Pt/SiO$_2$. The energies of the L absorption edges and the $L\alpha$ emissions are:

	L_3 (eV)	L_2 (eV)	L_1 (eV)	$L\alpha_{1,2}$ (eV)
Re	10535	11959	12527	8653,8586
Pt	11564	13273	13880	9442,9361
Au	11919	13734	14353	9713,9628

   A Which absorption edges would you choose to study for these two catalysts?

   B What path length would be viable for a transmission experiment (see Figure 6.15)?

   C What would be the viable k ranges for the EXAFS regions for these edges?

   D Which detection methods could you choose to extend the options for this study?

3 The carbide complex [(Bu^{t}O)$_3$W-C-C-W(OBu^t)$_3$] **C** (Figure 6.29) has a linear W-C-C-W unit with W-C and C-C distances of 1.82 and 1.34 Å in the crystal [Organometallics, 1992, **11**, 321–326]. In the alkoxy ligands, the W-O and O-C distances average at 1.447(1) and 1.51(1) Å, respectively, and the mean W-O-C angle is 142°.

Figure 6.29 Structure of complex C.

A Calculate the shells that you would expect to observe in a study of a solution of this complex using the W L_3 edge (10207 eV).

B Which of these shells would you expect to show significant multiple scattering effects?

C What is the maximum k range that you could record before reaching the L_2 edge (11544 eV)?

D What is the minimum k range that you should record to resolve all of these shells?

4   The Rh K edge XAFS of a 4 wt%Rh/Al$_2$O$_3$ sample is shown in Figure 6.30 as recorded under H$_2$, O$_2$, and NO, as part of a study of the oxidation of CO by NO.

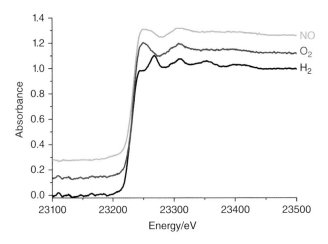

Figure 6.30   Rh K-edge XAFS of Rh/alumina in the presence of three gases.

A If you were to map the sites of oxidized and reduced rhodium in a catalyst bed, which energies would you adopt to make this estimation?

B What does the spectrum recorded under NO tell you about the average structure at rhodium?

C What additional measurements and analyses would you carry out to provide more detail about the reaction with NO?

References

1 'Extended X-ray absorption fine structure (EXAFS) characterization of the hydroformylation of oct-1-ene by dilute Rh-PEt₃ catalysts in supercritical carbon dioxide', S. G. Fiddy, J. Evans, T. Neisius, X.-Z. Sun, Z. Jie, M. W. George, Chem. Commun., 2004, 676–677.

2 'Insights in the mechanism of selective olefin oligomerization catalysis using stopped-flow freeze-quench techniques: A Mo K-edge QEXAFS study', S. A. Bartlett, P. P. Wells, M. Nachtegaal, A. J. Dent, G. Cibin, G. Reid, J. Evans, M. Tromp, J. Catal., 2011, **284**, 247–258.

3 'Activation of [CrCl₃{R-SN(H)S-R}] catalysts for selective trimerization of ethene: a freeze-quench Cr K-edge XAFS study', S. A. Bartlett, J. Moulin, M. Tromp, G. Reid, A. J. Dent, G. Cibin, D. S. McGuinness, J. Evans, ACS Catal., 2014, **4**, 4201–4204.

4 'Activation of [CrCl₃{PPh₂N(ⁱPr)PPh₂}] for the selective oligomerization of ethene: a Cr K-edge XAFS study', S. A. Bartlett, J. Moulin, M. Tromp, G. Reid, A. J. Dent, G. Cibin, D. S. McGuiness, J. Evans, Catal. Sci. Technol., 2016, **6**, 6237–6246.

5 'Detection of intermediate species in electron transfer between iron(III) nitrate and hydroquinone by millisecond time-resolved x-ray absorption spectroscopy in dispersive mode', N. Yoshida, T. Matsushita, S. Saigo, H. Oyanagi, H. Hashimoto, M. Fujimoto, Chem. Commun., 1990, 354–356.

6 'Time resolved stopped-flow x-ray absorption fine structure system using synchrotron radiation for fast reactions in solution', Y. Inada, H. Hayashi, S. Funahashi, M. Nomura, Rev. Sci. Instrum., 1997, **68**, 2973–2977.

7 'Active site electronic structure and dynamics during metalloenzyme catalysis', O. Kleifeld, A. Frenkel, J. M. L. Martin, I. Sagi, Nature Struct. Biol., 2003, **10**, 98–103.

8 'A new stand-alone QEXAFS data acquisition studies for *in situ* studies', J. Stötzel, D. Lützenkirchen-Hecht, R. Frahm, J. Synchrotron Radiat., 2011, **18**, 165–175.

9 'In situ XAFS experiments using a microfluidic cell: application to initial growth of CdSe nanocrystals', H. Oyanagi, Z. H. Sun, Y. Jiang, M. Uehara, H. Nakamura, K. Yamashita, L. Zhang, C. Lee, A. Fukano, H. Maeda, J. Synchrotron Radiat., 2011, **18**, 272–279.

10 'Application of stopped flow techniques and energy dispersive EXAFS for investigation of the reactions of transition metal complexes in solution: Activation of nickel β-diketonates to form homogeneous catalysts, electron transfer reactions involving iron(III) and oxidative addition to iridium(I)', M. B. B. Abdul Rahman, P. R. Bolton, J. Evans, A. J. Dent, I. Harvey, S. Diaz-Moreno, Faraday Discus., 2002, **122**, 211–222.

11 'Structural investigation of the bridged activated complex in the reaction between hexachloroiridate(IV) and pentacyanocobaltate(II)', S. Diaz-Moreno, D. T. Bowron, J. Evans, Dalton Trans., 2005, 3814–3817.

12 'Multitechnique approach to reveal the mechanism of copper(II)-catalyzed arylation reactions', M. Tromp, G. P. F. van Strijdonck, S. S. van Berkel, A. van den Hoogenband, M. C. Feiters, B. de Bruin, S. G. Fiddy, A. M. J. van der Eerden, J. A. van Bokhoven, P. W. N. M. van Leeuwen, D. C. Koningsberger, Organometallics, 2010, **29**, 3085–3097.

13 'Energy dispersive XAFS: Characterization of electronically excited states of copper(I) complexes', M. Tromp, A. J. Dent. J. Headspith, T. L. Easun, X.-Z. Sun, M. W. George, O. Mathon, G. Smolentsev, M. L. Hamilton, J. Evans, J. Phys. Chem. B, 2013, **117**, 7381–7387.

14 'Microsecond-resolved XAFS of the triplet excited-state of $Pt_2(P_2O_5H_2)_4^{4-}$', D. J. Thiel, P. Livins, E. A. Stern, A. Lewis, Nature, 1993, **362**, 40–43.

15 'MLCT state structure and dynamics of a copper(I) diimine complex characterized by pump-probe x-ray and laser spectroscopies and DFT calculations', L. X. Chen, G. B. Shaw, I. Novozhilova, T. Liu, G. Jennings, K. Attenkofer, G. J. Meyer, P. Coppens, J. Am. Chem. Soc., 2003, **125**, 7022–7034.

16 'Effects of electronic and nuclear interactions on the excited-state properties and structural dynamics of copper(I) diimine complexes', M. W. Mara, N. E. Jackson, J. Huang, A. B. Stickrath, X. Zhang, N. E. Gothard, M. A. Ratner, L. X. Chen, J. Phys. Chem. B, 2013, **117**, 1921–1931.

17 'A high-repetition rate scheme for synchrotron-based picosecond laser/x-ray probe experiments on chemical and biological systems in solution', F. A. Lima, C. J. Milne, D. C.V. Amarasinghe, M. H. Rittmann-Frank, R. M. van der Veen, M. Reinhard, V.-T. Pham, S. Karlsson, S. L. Johnson, D. Grolimund, C. Borca, T. Huthwelker, M. Janousch, F. van Mourik, R. Abela, M. Chergui, Rev. Sci. Instrum., 2011, **82**, 063111.

18 'Spatiotemporal stability of a femtosecond hard-X-ray undulator source studied by control of coherent optical phonons', P. Beaud, S. L. Johnson, A. Streun, R. Abela, D. Abramsohn, D. Grolimund, F. Krasniqi, T. Schmidt, V. Schlott, G. Ingold, Phys. Rev. Lett., 2007, **99**, 174801.

19 'Light-induced spin crossover in Fe(II)-based complexes: The full photocycle unraveled by ultrafast optical and x-ray spectroscopies', A. Cannizzo, C. J. Milne, C. Consani, W. Gawelda, C. Bressler, F. van Mourik, M. Chergui, Coord. Chem. Rev., 2010, **254**, 2677–2686.

20 'Femtosecond x-ray absorption spectroscopy at a hard x-ray free electron laser: application to spin crossover dynamics', H. T. Lemke, C. Bressler, L. X. Chen, D. M. Fritz, K. J. Gaffney, A. Galler, W. Gawelda, K. Haldrup, R. W. Hartsock, H. Ihee, J. Kim, K. H. Kim, J. H. Lee, M. M. Nielsen, A. B. Stickrath, W. Zhang, D. Zhu, M. Cammarata, J. Phys. Chem. A, 2013, **117**, 735–740.

21 'Ultrafast excited state relaxation of a metalloporphyrin revealed by femtosecond x-ray absorption spectroscopy', M. L. Shelby, P. J. Lestrange, N. E. Jackson, K. Haldrup, M. W. Mara, A. B. Stickrath, D. Zhu, H. T. Lemke, M.

Chollet, B. M. Hoffman, X. Li, L. X. Chen, J. Am. Chem. Soc., 2016, **138**, 8752–8764.

22 'The Mn_4Ca photosynthetic water-oxidation catalyst studied by simultaneous x-ray spectroscopy and crystallography using an x-ray free-electron laser', R. Tran, J. Kern, J. Hattne, S. Koroidov, J. Hellmich, R. Alonso-Mori, N. K. Sauter, U. Bergmann, J. Messinger, A. Zouni, J. Yano, V. K. Yachandra, Phil. Trans. R. Soc. B, 2014, **369**, 20130324.

23 'In situ XAS and XRPD parametric Rietveld refinement to understand dealumination of Y zeolite catalyst', G. Agostini, C. Lamberti, L. Palin, M. Milanesio, N. Danilina, B. Xu, M. Janousch, J. A. van Bokhoven, J. Am. Chem. Soc., 2010, **132**, 667–678.

24 'Identification of CO adsorption sites in supported Pt catalysts using high energy-resolution fluorescence detection x-ray spectroscopy', O. V. Safonova, M. Tromp, J. A. van Bokhoven, F. M. F. de Groot, J. Evans, P. Glatzel, J. Phys. Chem. B, 2006, **110**, 16162–16164.

25 'Susceptibility of a heterogeneous catalyst, Rh/γ-alumina, to rapid structural change by exposure to NO', T. Campbell, A. J. Dent, S. Diaz-Moreno, J. Evans, S. G. Fiddy, M. A. Newton, S. Turin, Chem. Commun., 2002, 304–305.

26 'Synchronous, time resolved, diffuse reflectance FTIR, energy dispersive EXAFS (EDE) and mass spectrometric investigation of the behavior of Rh catalysts during NO reduction by CO', M. A. Newton, B. Jyoti, A. J. Dent, S. G. Fiddy, J. Evans, Chem. Commun., 2004, 2382–2383.

27 'Time-resolved, *in situ* DRIFS/EDE/MS studies on alumina supported Rh catalysts: effects of ceriation on the Rh catalysts in the process of CO oxidation', A. B. Kroner, M. A. Newton, M. Tromp, A. E. Russell, A. J. Dent, J. Evans, Catal. Struct. Reactivity, 2017, **3**, 13–23.

28 'Fluorescence micro(spectro)scopy as a tool to study catalytic materials in action', G. De Cremer, B. F. Sels, D. E. De Vos, J. Hofkens, M. B. J. Roeffaers, Chem. Soc. Rev., 2010, **39**, 4703–4717.

29 'Visualizing a catalyst at work during the ignition of the catalytic partial oxidation of methane', B. Kimmerle, J.-D. Grunwaldt, A. Baiker, P. Glatzel, P. Boye, S. Stephan, C. G. Schroer, J. Phys. Chem. C, 2009, **113**, 3037–3040.

30 'Giant spin and orbital moment anisotropies of a Cu-phthalocyanine monolayer', S. Stepanow, A. Mugarza, G. Ceballos, P. Moras, J. C. Cezar, C. Carbone, P. Gambardella, Phys. Rev. B, 2010, **82**, 014405.

31 'XMCD proof of ferromagnetic behavior in ZnO nanoparticles', C. Guglieri, M. A. Laguna-Marco, M. A. García, N. Carmona, E. Céspedes, M. García-Hernández, A. Espinosa, J. Chaboy, J. Phys. Chem. C, 2012, **116**, 6608–6614.

32 'Characterizing magnetism of individual magnetosomes by x-ray magnetic circular dichroism in a scanning transmission x-ray microscope', K. P. Lam, A. P. Hitchcock, M. Obst, J. R. Lawrence, G. D. W. Swerhone, G. G. Leppard, T. Tyliszczak, C. Karunakaran, J. Wang, K. Kaznatcheev, D. A. Bazylinski, U. Lins, Chem. Geol., 2010, **270**, 110–116.

33 'Submicrometer hyperspectral x-ray imaging of heterogeneous rocks and geominerals: applications at the Fe *K*-edge', V. E. De Andrade, J. Susini, M. Salomé, O. Beraldin, C. Rigault, T. Heymes, E. Lewin, O. Vidal, Anal. Chem., 2011, **83**, 4220–4227.

34 'Analytical aspects of waterlogged wood in historical shipwrecks', Y. Fors, F. Jalilehvand, M. Sandström, Anal. Sci., 2011, **27**, 785–792.

35 'Sulfur and iron speciation in recently recovered timbers of the *Mary Rose* revealed via x-ray absorption spectroscopy', K. M. Wetherall, R. M. Moss, A. M. Jones, A. D. Smith, T. Skinner, D. M. Pickup, S. W. Goatham, A. V. Chadwick, R. J. Newport, J. Archeol. Sci., 2008, **35**, 1317–1328.

36 'Archeological applications of XAFS: Prehistoric paintings and medieval glasses', F. Farges, E. Chalmin, C. Vignaud, I. Pallot-Frossard, J. Susini, J. Barger, G. E. Brown Jr., M. Menu, Physica Scripta, 2005, **T115**, 885–887.

37 'Visualization of a lost painting by Vincent van Gogh using synchrotron radiation based x-ray fluorescence elemental mapping', J. Dik, K. Janssens, G. Van Der Snickt, L. van der Loeff, K. Rickers, M. Cotte, Anal. Chem., 2008, **80**, 6436–6442.

Index

Note: page numbers in italics refer to figures; page numbers in bold refer to tables.

X-Ray Absorption Spectroscopy for the Chemical and Materials Sciences, First Edition. John Evans.
© 2018 John Wiley & Sons Ltd. Published 2018 by John Wiley & Sons Ltd.